Numerical Investigation and Performance Enhancement of Scramjet Inlets and Combustors

Kathy Bano

TABLE OF CONTENTS

Chapter 3. SIMULATION METHODOLOGY

Chapter 4. NUMERICAL SIMULATIONS OF SCRAMJET INLET

Chapter 5. NUMERICAL SIMULATIONS OF SCRAMJET COMBUSTOR

CHAPTER 1

INTRODUCTION

1.1 Motivation

Since the Wright brothers' first powered flight, continuous progress in aviation and space propulsion has led to the evolution of gas turbines, rockets, and air-breathing engines. The power-to-weight ratio remains a critical parameter influencing overall performance. The need to develop hypersonic flight for rapid transportation and cost-effective access to space has produced a great deal of technological advancement. Gas turbine engines can operate only in the subsonic range while rockets produce high thrust but low specific impulse. The ramjet engine uses the forward motion of the vehicle as a means to compress air and mix fuel with it, there is no compressors or turbines required for function due to the aerodynamic design of the ramjet. The ramjet engine slows the incoming supersonic air to provide optimal conditions for burning fuel, when accelerated too much below Mach 7 too quickly becomes detrimental because of higher energy losses associated with a higher rate of skin friction and overheating of the ramjet surfaces.

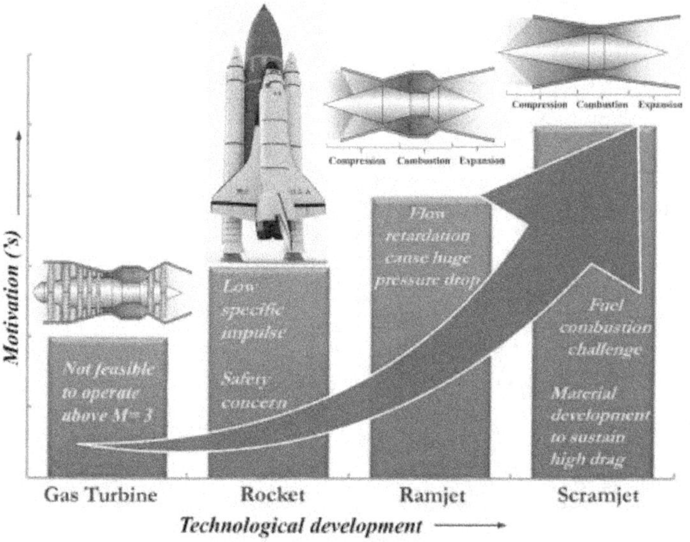

Fig 1.1: Evolution of Propulsion Technologies from Gas Turbine Engines to Scramjet systems [1]

Figure 1.1 demonstrates the developments of evolutionary technology from gas turbines engines to Scramjets. Technology in scramjet helps overcome the problem by sustaining the supersonic flow in the combustor, as opposed to ramjet engines in which the flow is slowed down. This technology has the potential to achieve hypersonic speeds. Other shortcomings are yet to be overcome. These include the completion of the combustion cycle in a short duration, the development of materials that are resilient to the effects of heat and drag, reliance on carrier vehicles in order to function, and the need for sophisticated laboratories where complicated combustion experiments have to be performed[2].

The Scramjet is a type of air-breathing jet propulsion system designed for hypersonic speeds, typically above Mach 5. However, unlike regular jet propulsion and ramjets, the Scramjet propulsion system does not incorporate any rotating equipment like compressors and turbines, respectively. In Scramjets, the propulsion system relies entirely on the moving vehicle compressing the air coming towards it before it enters the combustor. In this case, the high-speed air is ignited together with fuel, which is either hydrogen or hydrocarbons and then torch-fired and ignited while moving supersonically towards the back of the vehicle. The supersonic combustion produces the hypersonic force needed to propel the vehicle.

A scramjet differs from the traditional jet engines in many ways. On comparing the traditional turbojet engines or the turbofan engines with the scramjet, it has been realized that in the traditional engines, the airspeed slows down to subsonic speeds before the ignition, whereas the scramjet goes even further by maintaining the supersonic airspeed throughout the process, thus increasing the efficiency and power that results from this engine. The lack of moving parts or the simplicity involved will make the scramjet engines more reliable and lighter in weight, thus the requirements for the aerospace field.

The potential of the scramjet is revolutionarily transformative. Since the scramjets are able to function efficiently at high velocities and altitudes, the future of transport within near-space and the atmosphere may be revolutionized by the potential of the scramjet. The potential uses might be as simple as military reconnaissance missions or high-speed strike vehicles to the use as cost-satellite launch systems to hypersonic passenger vehicles that can go around the world within hours. The scramjets might additionally act as a seminal SSTO (single-stage-to-orbit) enabling technology between the gaps of flight at the atmospheric level to gaining access to the final frontier.

Nevertheless, though with great promise, scramjet technology is still in the developmental phase with many challenges that need to be resolved to achieve widespread use. Some of the most significant among these challenges are efficient injection and mixture of the fuel within the very brief residence time of the supersonic flow, stabilization of the combustion in the highly dynamic environment, thermal control to be able to withstand the high aerodynamic heat loading encountered at hypersonic velocities, as well as structural integrity retention and optimization of the engine performance over a very broad range of Mach numbers as challenges in both design as well as material engineering. Overall, scramjets are one of the most promising new frontiers to aerospace propulsion, yet there remains a great deal of research and technological development to be done. When such obstacles are resolved, scramjet-powered vehicles may transform aerospace engineering, introducing a new golden age of fast world transport as well as efficient space reach.

1.2 Air Breathing Engines and its types

An air-breathing jet engine is summed up as one of the most amazing feats of ingenuity in the aerospace engineering field, which was perfectly conceived to utilize atmospheric air for generation of the thrust. Through the intake section, the engine is supplied with massive amounts of air, which is the source of both energy and the feedstock for the release of the power, besides being the necessary oxidizer for combustion. After intake is the stage of compression, where the entering air is subjected to a fantastical change. Rotating compressor blades compress the air, vastly multiplying its pressure and temperature, fundamentally raising its density for future combustion.

After pressurization, the air–fuel mixture enters the combustion chamber, where ignition initiates combustion, producing high-temperature gases and further increasing the pressure. Continuing combustion, burning gases power turbines that are linked to the compressor, which converts thermal power into mechanical power. The output power is utilized in the compressor which is the part of the engine that takes air in circulation. And the remaining power is also used to propel the plane. The exhaust stage is the last one, with the expanded gases departing from the engine through a very accurately designed nozzle which provides the necessary push for the plane following Newton's third law.

Supersonic engines like Ramjets or Scramjets provide a giant leap in compression based on environmental requirements for efficient operation in supersonic speed. Ramjets function based on the concept that Ramjets are traveling along with the aircraft by compressing the incoming air by the force of thrust up to Mach 5. Scramjets operate in hypersonic speed up to Mach 15 based on the high compression and heating technique of air. Supersonic engines bring tremendous changes in the field of transport, power, and space travel by lifting man into new horizons. The compression, combustion, and expansion of different air breathing jet engines is shown below in Figure 1.2

A scramjet is a new type of engine with many benefits regarding the working of aircraft during high speeds. Since it is capable of holding its speed during hypersonic speed, a scramjet is more reliable for hypersonic cruise parameters. The airflow is assured to work at a supersonic speed, which is a result of the technological development of the inlet and the scramjet engine's pressure-generating combustors. The Combustor ensures a sufficient supply of air, fuel, and airflow for an improved combustion process beforehand leading to the nozzle, where it is effortlessly mixed for the generation of an effective force of thrust. It still has significant contributions to make regarding the aliment, of hypersonic speed by employing an advanced method of air and fuel combustion. The stability of an advanced generation of scramjets is dependent on the inlet of a scramjet.

After passing through the inlet at hypersonic speeds, air first slows down and is compressed before it enters the combustion chamber, where fuel is injected and burning at a supersonic speed takes place. The air then passes through a nozzle, where thrust is produced. The complexities of the air flow and the shockwave interaction that happen in an inlet have to be carefully analyzed if one desires to increase the performance level of the engine.

To sum up, scramjets are sophisticated and complicated super cool engines, which operate in extreme conditions; hence they need the flawlessness of advanced engineering and materials to operate properly. The technology has the potential of delivering high-speed travel within the atmosphere and could potentially be developed for use in both military and civilian transports.

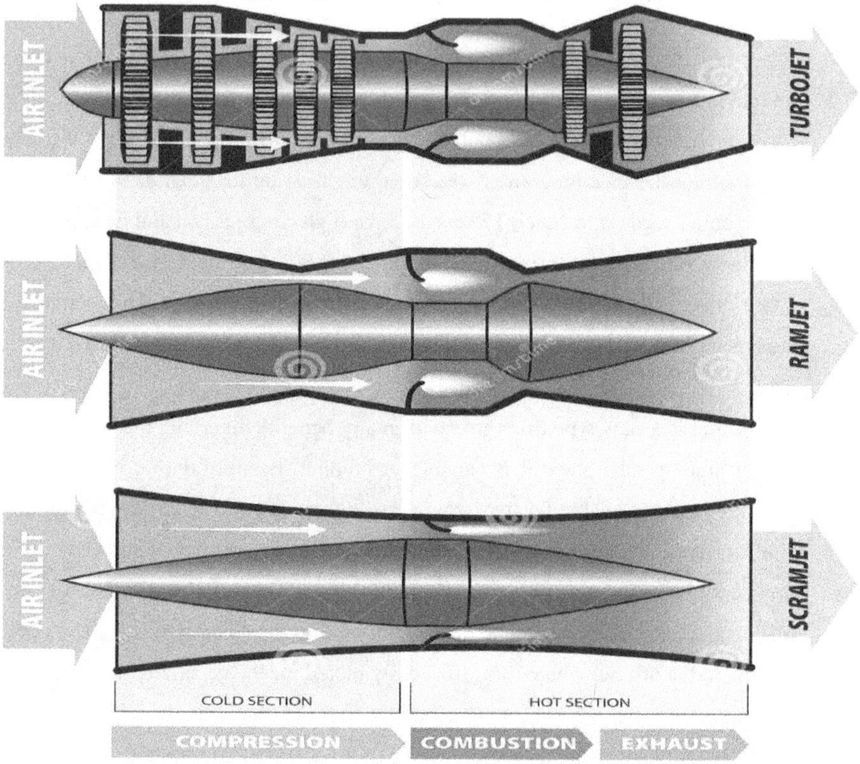

Figure 1.2: The compression, combustion, and expansion regions of (a) turbojet, (b) ramjet, and (c) scramjet engines [1].

1.3 Scramjets

The supersonic combustion ramjet engine or scramjet utilizes air compression through hypersonic flight speeds (Mach 5+) without the need of moving parts. This makes them highly effective for a variety of aerospace applications such as long-range missile systems or spaceplanes. Scramjets rely only on the natural compression and ignited fuel mixtures created by the engine for its thrust, making the engine extremely lightweight.

Utilizing a supersonic mix of air entering the engine and fuel stored inside, scramjets utilize a simple yet elegant design to provide new avenues of success for developing extremely fast and efficient propulsion mechanisms within aerospace. The working mechanism of the Scramjets is shown below in Figure 1.3. The focus of the continual development of scramjet propulsion technologies will continue to lead to providing an avenue to develop even higher velocity and more efficient engines and thrust sources for aerospace applications and beyond.

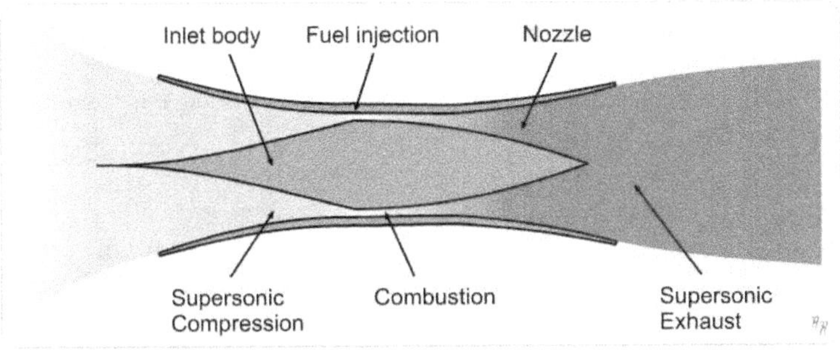

Figure 1.3 Scramjet working [1]

1.4 Scramjet Components

An engine that utilizes scramjet technology uses three primary components for hypersonic propulsion: Inlet, combustion chamber/combustor and nozzle. Each component is essential for providing efficient thrust at high speeds.

The inlet collects air coming from the atmosphere, compresses it and cools it down before sending it to the combustion chamber. This process does not use mechanical devices such as fans or pumps; instead, the shape of the inlet uses aerodynamic forces to change some of the dynamic pressure of the oncoming flow into static pressure. As a result, the pressure and temperature of the air increase to acceptable levels for allowing combustion.

The combustor mixes the compressed air with hydrogen (the most common fuel used) in the supersonic combustion chamber. The combustion process in this environment must happen very quickly because the total time available for the air to remain in the combustion chamber is only milliseconds. Efficient mixing of fuel and air, ignition, stabilization and heat transfer can be accomplished by using specially designed injectors and flame-holding devices and creating shock waves.

Following the combustion chamber is the nozzle; it expands, accelerates and transforms the high temperature combustion gases into thrust by converting the thermal energy of the combustion products into kinetic energy. The shape of the nozzle is optimized to create supersonic and in some cases hypersonic flow velocities in order to maximize propulsive efficiency.

1.5 Working of Scramjets

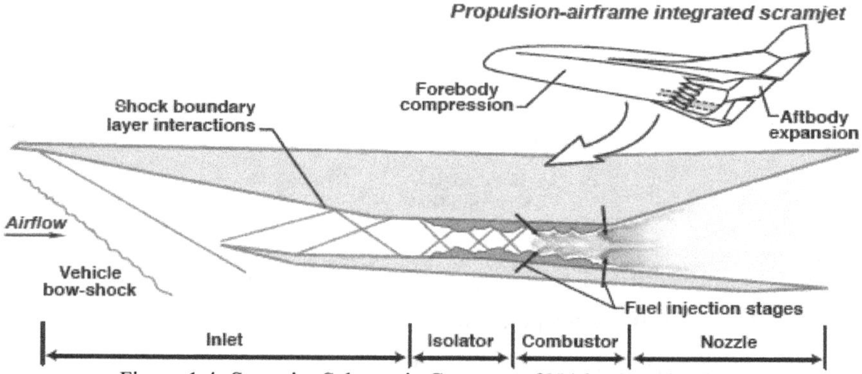

Figure 1.4: Scramjet Schematic Courtesy of NASA Langley [1]

Scramjets, or Supersonic Combustion Ramjets, are a type of engine that uses air-breathing propulsion to achieve hypersonic speeds (speeds greater than Mach 5). Scramjets differ from traditional jet engines in that they do not have rotating parts like turbines, and they operate at much higher speeds as shown in Figure 1.4.

A scramjet's operation relies on compressing air as it enters the engine, which is accomplished by the aircraft's own speed. A supersonic combustion process is produced when heated and compressed air is injected into the flow and ignited. The high- speed airflow through the engine maintains this combustion process and generates a thrust that propels the aircraft forward[1-4].

Scramjets are designed to operate at very high altitudes and speeds, where the air is thin and the traditional jet engines cannot operate efficiently. The faster the aircraft travels, the more efficient the scramjet becomes, as the high-speed airflow through the engine increases the combustion efficiency. Overall, scramjets are incredibly sophisticated and complex engines that need to be engineered and made with very specific materials in order to fly at such extreme temperatures. They hold the promise of atmospheric high-speed travel, and could potentially be used on future aircraft and space vehicles for a variety of purposes, such as military and civilian transportation.

1.6 Scramjet Engine Performance Parameters

The scramjet engine performs autonomously based on the efficient utilization of its main constituents the inlet, combustor, and nozzle. Each is responsible for uniquely contributing to the overall propulsion characteristics and can be measured quantitatively with the aid of a number of parameters.

1.6.1 Inlet Performance Parameters

• **Mach Number:** The scramjet operates with the inlet Mach number dictating the operating mode (hypersonic or supersonic). The inlet Mach number is the main factor that changes the air compression, the creation of the shock, and the recovery pressure in general. A scramjet is basically a device that works most efficiently in a Mach 5–10 speed range.

• **Thermodynamic Efficiency:** This is the ability of the inlet to compress the ingested flow with minimum loss of energies. The higher the thermodynamic efficiency, the better inlet will be able to accelerate and obtain the overall required pressure and temperature for combustion with minimal increase in entropy or drag loss.

• **Kinetic Energy Efficiency:** This efficiency will indicate the inlet can effectively convert the entering kinetic energy of the high-speed flow into pressure energy. They indicate the quality of the compression of shocks as well as the smoothness of the flow path, both of which are important factors in minimizing total pressure loss.

• **Total Pressure Recovery Ratio:** Provided as the ratio of the combustor inlet total pressure to the freestream one, the variable accounts as the indication of the loss of pressure within the inlet. The higher the recovery ratio, it is the outcome of lower shock as well as friction loss correspondingly and thus presents a good performance of the engine together with the enhanced thrust generation.

1.6.2 Combustor Performance Parameters

- **Specific Thrust:** The net thrust generated per unit mass flow rate of air is referred to as specific thrust. It depends on combustion efficiency, fuel-air mixing, and expansion through the nozzle. In scramjets, maintaining high specific thrust at hypersonic speeds is a key design objective.

- **Specific Impulse (Isp):** Specific impulse is the thrust resulting from the unit flow rate of fuel weight and is considered as a measure of propulsive efficiency. The higher the Isp, the more the engine is able to get useful thrust from each kilogram of fuel. Generally, scramjets have higher Isp than rockets since they use air as an oxidizer instead of carrying oxidizers in the vehicle.
- **Combustion Equivalency Ratio (ϕ):** This ratio contrasts the stoichiometric and actual fuel-air ratios. It shows if the mixture is rich ($\phi > 1$), stoichiometric ($\phi = 1$), or lean ($\phi < 1$). Proper control of the equivalency ratio is essential for stable combustion and optimal thermal efficiency under supersonic conditions.
- **Species Molecular Weight:** The average molecular weight of the combustion products affects exhaust velocity and, consequently, thrust. Lower molecular weight species (e.g., hydrogen combustion products) yield higher exhaust velocities, improving engine performance.
- **Maximum Mixing Efficiency:** Mixing efficiency represents how well the injected fuel mixes with the supersonic airstream before ignition. One of the main difficulties in designing scramjet combustion is achieving high mixing efficiency in a very short residence time. Efficient mixing ensures uniform combustion, reduced pressure losses, and improved thrust generation.

Table 1.1: Summary of Scramjet Engine Performance Parameters

Component	Parameter	Definition / Description	Influence on Performance
Inlet	Mach Number	Ratio of flight velocity to local speed of sound; defines the operating regime (supersonic/hypersonic).	Determines inlet compression level, shock structure, and engine efficiency.
	Thermodynamic Efficiency	Measures how effectively the inlet	Higher efficiency improves total pressure and

		compresses air with minimal entropy and energy losses.	temperature rise, enhancing combustion efficiency.
	Kinetic Energy Efficiency	Fraction of incoming kinetic energy converted into useful pressure energy.	High values indicate smooth compression and reduced drag losses, improving overall inlet performance.
	Total Pressure Recovery Ratio	Ratio of total pressure at combustor entry to freestream total pressure.	Indicates inlet pressure losses due to shocks and friction; higher ratio means better engine performance.
Combustor	Specific Thrust	Net thrust produced per unit mass flow rate of air.	High specific thrust leads to better propulsion efficiency at hypersonic speeds.
	Specific Impulse (Isp)	Thrust produced per unit fuel weight flow rate.	Higher Isp signifies greater propulsive efficiency and effective fuel utilization.
	Combustion Equivalency Ratio (ϕ)	Ratio of actual fuel, air ratio to stoichiometric ratio.	Controls combustion stability, flame temperature, and efficiency; optimal ϕ ensures stable operation.
	Species Molecular Weight	Average molecular weight of combustion products.	Lower molecular weight yields higher exhaust velocity and improved thrust.
	Maximum Mixing Efficiency	Degree to which injected fuel mixes with supersonic air before ignition.	High mixing efficiency ensures uniform combustion, reduced losses, and higher thrust output.

1.7 Global Trends in Scramjet Technology

Worldwide scramjet technology research has mainly depended on the contributions of few leading countries, each having a different driver for the development of hypersonic propulsion. One of the biggest contributions comes from the US where the ground and the flight experiments on the HyTech and Hyper-X (X-43) projects helped a lot in the progress of a liquid hydrocarbon fueled scramjet, refinement of the inlet and isolator design, and combustion performance improvement. X-43A flight at high altitudes and high speeds was able to break the record of more than Mach 9.6, thus proving that an air-breathing engine is indeed feasible in the hypersonic regime. After the HyTech project had resolved the problems related to structural control of the vehicle, thermal regulation, and even the integration of the vehicle in the fuel circuit, the serpentine inlet design was able to show its efficiency in the field of both aerodynamics and stealth technology for the future of space missions [5, 28-30].

On the other hand, Germany plays a major role in conducting experimental and numerical research, which is a project led by the German Aerospace Center (DLR). Using high-speed wind tunnels such as the H2K facility, Germany has thoroughly investigated Mach 6 inlet performance, shock-boundary layer interactions, and passive boundary-layer bleeding with a view toward maintaining flow stability and preventing unstart. These experiments, supported by the German Research Foundation (DFG), have provided important data for validation of CFD models and insight into inlet aerothermodynamics at hypersonic conditions [13-14, 22].

In addition, Russia has one of the longest and most comprehensive programs of scramjet technology research tracing back to the 1950s when CIAM initiated this program. Among the various experiments is the flight tests of Kholod, which demonstrated hydrogen-powered scramjet operation at speeds up to Mach 6.5, and IGLA, up to Mach 16. The contribution of Russia to works on dual-mode engines, pseudo shock phenomena, and supersonic combustion has established the basis of global scramjet development and design practice [5].

Moreover, China has been very swift to catch up with the hypersonic field of technology, especially with the production of the DF-ZF and Xingkong-2 as the main use of both military and civilian. These undertakings have enabled China to design its own inlets, optimize its combustion systems, and research advanced materials. Among these are technically operational hypersonic glide vehicles and cruise missiles like the CM-400AKG, which has made China one of the top-tiered countries in hypersonics technology, but rather, as a consequence, it escalates regional security and the proliferation of nuclear weapons concerns[49].

Long-term programs like ESOPE, PREPHA, and PROMETHEE have been France's contribution in the fields of hydrogen-fueled scramjets, strut injector analysis based on CFD, and variable-geometry inlet design, respectively. Cooperative efforts such as Wide Range Ramjet (WRR) with Russia have moved research along engines operative between Mach 3 to 12, indicating continued interest by France in useful scramjet propulsion systems [5].

Japan has gained significant success with the NAL Kakuda Research Centre, where testing of hydrogen-fueled scramjets has been conducted in a Mach number regime from 4 up to 15 in facilities such as the HEIST shock tunnel. The Japanese have contributed extensively to knowledge on sidewall compression inlets, shock-generating ramps, and strut injectors for substantially improving knowledge on combustion efficiency, flame stability, and heat transfer issues in hypersonic environments [5].

The UK has participated in various international collaborations such as the HyShot 2 flight experiment with Australia and the LAPCAT projects with the European Union. These have contributed to inlet compression theory and data analysis, together with work on developing combined-cycle propulsion systems including the SABRE engine concept for reusable spaceplanes [5,6].

Thus, India has been moving fast to become a hypersonic power with its HSTDV and BrahMos II. A successful Mach 6 scramjet test flight was carried out by India's DRDO in 2020, which established indigenous capability in high-speed propulsion and underlined its strategic cooperation with Russia in developing next-generation hypersonic systems[1,5].

The European Union, via LAPCAT I and II, has funded and managed multi-national collaborations that focus on the design of environmentally-friendly Mach 5–8 hypersonic aircraft that include the integration of airframe-propulsion systems, CFD validation, and research on the environmental impact. These initiatives stand as Europe's consolidated

1.8 Layout of thesis

The report comprises seven chapters. In each chapter, parts of the dissertation are included as follows:

Chapter I Introduction: This chapter provides an overview of the working principles of the Scramjet propulsion systems. The chapter also discusses the reason why the present investigation is carried out which is largely attributed to technological gaps and the need for efficient designs of the inlet and combustion system. The chapter also analyzes the scramjet technology trends worldwide.

Chapter II Literature Review: The chapter is aimed at providing detailed reviews of relevant works related to the development of the scramjet. Also, the chapter has been instrumental in the understanding of tools and methods necessary to carry out the proposed research work. This chapter ends with the identification of the critical gaps in research that led to this investigation. The chapter also outlines the problem statement and defines the study objectives and scope.

Chapter III Numerical Methodology: The chapter describes the study computational approach in detail. It deals with the fluid dynamic theory and the numerical aspects of CFD used in the analysis of a Scramjet.

Chapter IV Numerical Simulation of Scramjet Inlet: This chapter presents the numerical analysis of the Inlet of a Scramjet. It describes the design methodology, the validation, and the techniques for the performance enhancement of the inlet to the optimization of its performance.

Chapter V Numerical Simulation of Scramjet Combustor: The chapter demonstrates the computational study of the design and operation of the scramjet combustor. It includes the evaluation of the performance under different geometric and operational conditions.

Chapter VI Integrated Analysis of Scramjet (Inlet + Combustor): The chapter deals with the integration of lead configurations of the chamber and the inlet of the Scramjet into a single model to assess the overall performance of the engine.

Chapter VII Conclusions and Future Work: The last chapter presents the summary of the principal results of the study and highlights the most important findings. It lays out the significant conclusions drawn from the research and offers possible future research directions that could be pursued.

CHAPTER 2

LITERATURE REVIEW

2.1 Introduction

The effect of different variables on shock-induced flow separation in supersonic and hypersonic flows has been the main area of aerodynamic research for more than 50 years. The first researches, mainly in the 1960s and 1970s, contributed significantly to the present knowledge of the issue. These studies examined how various factors such as shock strength, boundary-layer thickness, surface geometry, and pressure gradients affect the arrival and the degree of flow separation after the shock interactions. Since Computational Fluid Dynamics (CFD) tools were not advanced at that time, a great deal of literature from that era was based on the experimental method only, which included wind tunnel testing, pressure measurements, and flow visualization techniques such as Schlieren photography and oil-flow patterns. These experiments allowed for the creation of empirical correlations and provided significant insight into the complicated nature of Shock Boundary Layer Interactions (SBLI), which are extremely important in determining the aerodynamic performance, stability, and heat transfer characteristics of high-speed vehicles [2].

Hypersonic air-breathing vehicles have been the focus of the global research during the last two decades, and this has been the main reason why the interest in shock-induced separation has been revived. The progress in computational power, turbulence modeling, and simulation techniques has allowed researchers to re-examine the classical research issues with much more accuracy and understanding. During the last 20 years, the majority of the studies have been using Reynolds-Averaged Navier–Stokes (RANS), Large Eddy Simulation (LES), and Direct Numerical Simulation (DNS) to understand the detailed physics of SBLI at realistic Mach number, Reynolds number, and thermal conditions. These numerical experiments complement the physical ones, giving a deeper understanding of the unsteady nature of shock motion, separation bubble dynamics, and their influence on aerothermal loads and structural integrity in scramjet inlets and other high-speed flow systems [5].

Thanks to the perfect combination of present CFD simulations and advanced diagnostic techniques, there is now a much clearer understanding of the fundamental mechanisms of shock-induced separation, thus design optimization of inlets, isolators, and combustors in hypersonic engines has reached a higher level of sophistication. Hence, present-day studies not only rely on the 20th-century experimental groundwork but also have the benefit of numerical accuracy and interdisciplinary modeling to address the challenges of future hypersonic flight [1].

A comprehensive review of scramjet engine research involves various critical aspects such as design, analysis, computational study, and experimental validation. The goal of such a review is to connect the theoretical basis of supersonic combustion with the practical aspects necessary for high-speed air-breathing propulsion. The design stage usually consists of the optimization of inlet, isolator, and combustor configurations for the purposes of efficient air compression, stable combustion, and effective thrust generation at different Mach numbers. Analytical and numerical studies, which are mostly done through Computational Fluid Dynamics (CFD), provide researchers with tools to create the model of the complex flow phenomena such as shock–boundary layer interactions, thermal loads, mixing efficiency, and flame stabilization in the combustor.

Experiments have also an important role in this field as they provide the proof for the computational models and give the real-world insight of the aerothermal behavior of scramjet components. Besides, the tests in hypersonic wind tunnels, shock tunnels, and free-jet facilities are the means for the flow structures, ignition characteristics, and material performance observation under the most extreme conditions. These activities jointly lead to a thorough understanding of scramjet operation, thus covering the gap between theory and practical implementation.

2.2 Literature Review on Inlets

Over the years, the aerospace propulsion sector has advanced from the initial subsonic systems to the development of air breathing engine systems that are both fast and efficient at the same time. Among these engines is the scramjet, which is now regarded as a revolutionary hypersonic flight technology. It is a supersonic combustion ramjet which can operate for a long time at speeds beyond Mach 5. Conventional gas turbine engines differ from scramjet engines in that the latter do not have rotating parts but instead rely solely on the vehicle's motion to compress the incoming air. This unique feature turns the inlet into a vital subsystem that slows down and compresses the superfast air stream and prepares it for burning [2, 3]. Ramjet engines have inlets that slow down and compress the air to subsonic speeds before combustion; on the other hand, scramjet combustion takes place in a supersonic flow which requires that the inlet be able to compress the high-Mach freestream effectively with minimum total pressure losses and also without flow separation [4]. Combustion stability and start-up capability are closely linked to the inlet design effectiveness for a large variety of flight conditions. As the critical element to the management of these phenomena, the inlet is still the subject of various design strategies: external, internal, and mixed compression schemes. The performance envelope of scramjets has also been greatly expanded with the help of inlet geometry innovation, variable-configuration designs, and boundary-layer control [6-8].

The degradation trends at the lower contraction ratio conditions have been obtained in other experimental inlet investigations of 2D and 3D configurations, where separation and shock boundary-layer interactions reduce the effective mass capture [11-15]. The use of internal contraction as an enabling strategy for achieving high mass-flow utilization is further reinforced by emerging variable-geometry inlet concepts. Movable cowl and sliding panel configurations researched by Hohn & Gülhan [12] and Ogawa et al. [17] show that varying the internal contraction dynamically can keep the mass-capture ratio and pressure recovery local over a wider Mach number and angle of attack range. Scramjet intake (inlet) structures might be classified in a very general way such as geometrical-wise, nature of compression-wise, and application-wise. On the geometrical side, they may be 2D (two-dimensional), 3D (three-dimensional), or axisymmetric. Usually, 2D inlet uses planar ramps, and due to its simple nature, it is easier to analyze and manufacture; hence, it is mainly used in experimental and conceptual studies. Meanwhile, 3D inlets, which may have some sidewalls or a contoured surface, can offer more mass capture as well as higher pressure recovery by alleviating the crossflow effects [61-65]. Axisymmetric inlets are the ones generally employed in missile-like configurations or high-speed flight vehicles, where the requirements are for a uniform flow and a compact design[17-20].

According to compression features, scramjet inlets are categorized into external, internal,

The Kantrowitz limit illustrated in Equation (1) represents the maximum theoretical mass flow rate through an inlet at which choking occurs. It is a function of the Mach number and the specific heat ratio (γ). Beyond this limit, flow choking and unstart may occur, thus it is a very important factor in the design of efficient and stable scramjet operation. [26-27].

The Stanton number is calculated as follows [18],

$$C_p = \frac{p - p_\infty}{0.5 \rho_\infty V_\infty^2} \tag{2}$$

$$T_{\text{rec}} = (1 + r\frac{\gamma - 1}{2} M_\infty^2) T_\infty \tag{3}$$

$$St = \frac{\dot{q}_{conv}}{\rho_\infty V_\infty C_{p,\text{air}} (T_{\text{rec}} - T_w)} \tag{4}$$

Usually, internal compression inlets are upgraded with some flow control elements like boundary-layer bleeds or vortex generators to quiet separation and keep attached flow over the ramps. The main outcome of such a process is a more stable and reliable inlet flow field, which is absolutely necessary under the quickly changing conditions of hypersonic flight. So, internal compression methods, along with the support of experimental and computational studies [9,12,18-23], have significantly contributed to pressure recovery and inlet reliability. Besides, variable-geometry designs and flow control techniques, they are the means of modern scramjet engines to be able to get stable start/unstart working and high performance over a wide flight envelope.

2.2.3 Mixed Compression

Mixed-compression inlets use both external and internal compression to balance performance, stability, and thermal loading. The forebody and ramps give the initial compression, and then more contraction is done inside the inlet. This setup is supposed to reduce total pressure losses and increase start ability under different conditions.

Matthews et al. [10] carried out an experimental verification of the isentropic-spike (re-expansion) intake at Mach 7.1, demonstrating that externally and internally coordinated compression results in high performance with a low risk of unstart (Fig.2.2). Nevertheless, there are still some problems, such as maintaining uniform flow at the isolator entrance and achieving correct shock balancing for various flight conditions. Improper geometric tuning can result in separation or drag, thus the problems that arise from pure external or internal configurations are reproduced. [11,23-25].

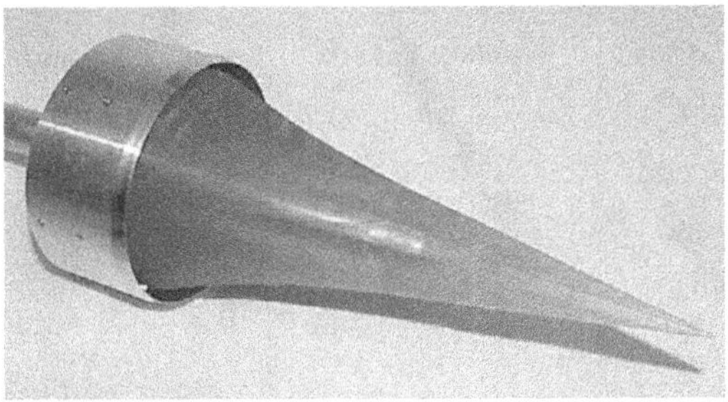

Fig. 2.2: Axisymmetric Intake Model [10].

Despite these difficulties, when properly designed and balanced, mixed-compression inlets offer robust starting characteristics, a wide operating range, and high pressure recovery, making them an ideal compromise.

Table 2.1: Comparison of Scramjet Inlet types.

Aspect	**External Compression** [9,23]	**Internal Compression** [11, 12]	**Mixed Compression** [10]
Compression Location	• Occurs mainly on forebody ramps (before cowl)	• Takes place inside the inlet duct (after cowl lip)	• Combines both external and internal compression zones
Operational Mach Range	• Typically, Mach 6–8+ • Narrow off-design tolerance	• Suitable for Mach 6–8+ • Supports high contraction ratios	• Broad Mach range if balanced properly • Requires tuning
Pressure Recovery	• Often reduced due to shock-induced separation and spillage	• High recovery due to better shock–boundary-layer control	• High recovery possible if shock systems are well-matched
Start ability	• Sensitive to shock-induced separation • Poor unstart margin	• Stable starting at higher contraction ratios	• Depends on shock balance and geometry • May need validation

Flow Separation	• Frequent due to sidewall shocks and vortex formation	• Less prone to separation; more attached flow	• Risk depends on external/internal balance
Mass Capture Ratio	• Reduced at off-design conditions	• Better retention over varying conditions	• Intermediate, based on flow uniformity and spillage control
Thermal Loads	• High at external ramps and leading edges	• Concentrated internally; better for cooling management	• Distributed between forebody and duct surfaces
Structural Complexity	• Simpler ramps; less internal detail	• More complex internal geometry (ramps, contraction ducts)	• Most complex; requires precise integration
Off-Design Sensitivity	• High; prone to unstart and performance drop	• Lower; more robust under varying conditions	• Moderate; sensitive to both regimes
Examples	• 2-ramp inlets with sidewall compression (DLR H2K)	• Movable cowls with internal compression (DLR H2K)	• Isentropic spike intakes (REX1020, Mach 7.1 tests)
Strengths	• Simpler integration • Initial compression outside engine	• Greater start margin • Higher efficiency • Robust flow	• Combines efficiency with design flexibility
Weaknesses	• Poor startability • Prone to separation and spillage	• More complex to design/manufacture	• Difficult to optimize • Can suffer from both inlet drawbacks

In order to reveal the flow properties and to create new methods of performance evaluation the team led by Azam Che Idris et al. [31] tested pressure-sensitive paint as a non-intrusive technique to study scramjet inlet fluid mechanics at Mach 5. Their research was significantly centered on the luminescent measurement systems, through which they explored a Scramjet Inlet-Isolator. The results indicated that an angle of attack of 2°

works to the advantage of performance as it lessens the flow pressure at a more considerable efficiency of the isolator exit. However, the enhancement by a further 2° is only marginal, and a simultaneous drop in pressure and energy is observed. Besides, the removal of Mach stem in the case of AoA=4° can increase the total performance.

When speaking about scramjet Seong-kyunIm et al. [32] argued that for safe flights it is vital to understand unstart due to flow choking which also comprises of the knowledge of the choking mechanism, diagnostic indicators, and control both in open and closed loop. In their performance of the time-accurate pressure measurements in the dual-mode ramjet/scramjet inlet, Patrick E. Rodi et al. [33] were able to infer that unstart caused by cowl-induced leads to oscillatory behavior at frequencies centered around 300 Hz, contraction ratio of the inlet influences pressure measurements profoundly and pressure increases if backed by inlet are allowed to increase further.

Michael K. Smart [34] pointed out the decisive role of scramjet inlets in hypersonic engines and noted that their design is the most significant factor in the overall propulsion system performance. They recommend the use of combustion chamber pressure to set compression levels and point out the three-dimensional inlet configurations as a promising solution for efficient flight at Mach 6 to 12. M. K. Smart [35] demonstrated that the conditions for maximum total pressure recovery in a two-dimensional scramjet inlet also imply nearly equal strength shock waves, even with additional constraints imposed by scramjet requirements, and that Lagrange multipliers present an effective tool in this case for optimization.

Hideaki Ogawa et al. [36] performed multi-objective design optimization for axisymmetric scramjet inlets, uncovering the effects of different variables such as inlet exit radius and showing by example the high-performance benefits of Busemann-type geometries. Suresh P et al. [37] through numerical optimization efforts in supersonic inlet designs aimed at total pressure recovery maximization and shock interaction minimization while also validating their proposal methods through theoretical data and CFD simulations. M.C. Neale [38] noted that a two-dimensional combined external/internal compression intake had reached a pressure recovery of 87% at Mach 2.2, with a possibility of further raising the figure up to 90%, thus stressing the significance of bleed slot design and subsonic diffusion efficiency.

investigations that introducing slight cowl deflection angles to a mixed compression air-intake designed for Mach 2·2 leads to improved performance and starting behavior, comparable to traditional bleeding methods.

Levi Reuben Vr et al. [53] optimized a 2D mixed-compression supersonic intake at Mach 2.4, finding that a 5-degree cowl deflection maximizes total pressure recovery, ensuring effective compression of supersonic flow. Derek J. Dalle et al. [54] investigated variable-geometry cowl designs for two-dimensional scramjet inlets, aiming to enhance performance across various Mach numbers. N K Gahlot et al. [55] analyzed the starting behavior of a supersonic air intake using computational simulations with cowl porosity, revealing enhanced intake performance and flow quality. Ved Merchant et al. [56] applied methods such as adjusting cowl deflection angles and bleed deflection angles to boost efficiency in a Mach 2.2 ramjet inlet.

B. John et al. [57] explored enhancing a supersonic air intake model's performance with blunted leading edges, revealing improved mass capture and combustion stability alongside reduced shock wave boundary layer interaction and flow distortion, albeit with a slight decrease in total pressure recovery. An optimised two-dimensional mixed compression scramjet inlet design for spontaneous hydrogen combustion at supersonic speeds was presented by Pedro P.B. Araújo et al. [58]. This design showed successful combustion chamber entry conditions and consistent thermodynamic properties across a range of flight Mach numbers. P R Viswanath [59] emphasizes the perennial difficulty in the turbulence interaction and Reynolds number variations in the interaction of shock-wave-boundary-layer, insisting that additional research be conducted to reduce such complexities.

Table 2.2: Comparative Summary of Scramjet Inlet Configurations

Aspect	**2D Inlet**	**3D Inlet**	**Variable Geometry Inlet**	**Axisymmetric Inlet**
Geometry	• Planar ramps • No sidewall flow [11]	• Includes sidewalls • Produces spanwise non-	• Movable cowl/ramp adjusts contraction ratio [12], [17]	• Central spike or cone geometry for circular bodies

		uniformities [11], [12], [23]		[10]
Start ability	• Sensitive to cowl-shock separation • Limited Mach range [11]	• Influenced by sidewall geometry• Sensitive to angle of attack and yaw [12], [23]	• Enhanced unstart margin across flight envelope [12], [66,69-70]	• Requires specialized methods (e.g., diaphragms, doors) [17]
Flow Stability	• Unsteady at off-design [11]	• Prone to vortex formation and corner separations [9], [23]	• Adaptive geometry allows shock re-positioning [12], [17]	• Susceptible to swirl distortion and heat gradients [10]
Mass Capture	• Lower due to 2D profile limitations [11]	• Higher due to better control of 3D shock system [12], [18]	• Actively managed through geometric tuning [12]	• High but depends on centerline symmetry [10]
Thermal Load Managem ent	• Predictable but concentrated near ramps [11]	• High heat flux at sidewalls and corners [13], [14]	• Distribution controlled via real-time adjustment [17]	• Requires advanced materials near spike and cone [10]
Flow Control Options	• Passive bleed to mitigate separation [53], [14]	• Internal/externa l bleed, corner inserts [12], [18]	• Integrated sensors/actuators with closed-loop control [17]	• Stream-thrust averaging, strong-weak shock management [10], [30]
Design Complexi ty	• Simple for analysis and wind tunnel testing [11]	• Complex to manufacture and simulate [9], [12]	• High mechanical/control system complexity [17]	• Moderate, but sensitive to symmetry and spike design [10]
Use Cases	• Baseline experiments	• Realistic representation	• Advanced adaptive flight scenarios [12],	• Axisymmetric vehicles

	and CFD validation [14]	of vehicle intakes [12], [18], [23]	[17]	(missiles, cones, blended bodies) [10]
Key Advantages	• Simple structure • Well-documented behavior [11]	• Enhanced recovery and adaptability [12], [18]	• Robust startability and shock control [12], [17]	• High integration efficiency • Ideal for round-body designs [20]
Key Challenges	• Low compression ratio • Poor off-design performance [11]	• Sensitive to off-design angles and flow asymmetries [12], [23]	• Mechanical reliability and real-time control complexity [17]	• Complex start-up dynamics • Potential swirl distortion [10]

2.3 Research Gaps in Inlet:

The existing research condition on scramjet engines shows a big gap how the integration of bleed of various dimensions as well as cowl deflection of various angles will impact the scramjet inlet performance.

Objective: The current research is done with different bleed dimensions and different cowl angle deflection one at a time and then the focus was made on what will be the scramjet inlet performance effect because of the integration of cowl angle with bleed.

The current research attempts to bridge this gap by performing a thorough computation analysis of the Mach 5 scramjets inlet, with particular regard to two-dimensional scramjet inlet-isolator geometry. Recognizing how paramount scramjets are to the development of aerospace engineering, this research hopes to enhance the performance of scramjet engine inlets.

Problems such as shoulder separation and intake system performance optimization were resolved using an passive flow control method as well as geometric alterations of scramjet inlets. The impact of passive boundary layer bleed at the inlet throat on flow parameters with regard to different bleed dimensions is specifically examined in this study.

The aim of this work is to obtain a broader understanding of the mechanisms that control scramjet inlet performance based on the fact that a previously studied 2-D inlet design has already produced some results. In the end, acquiring the knowledge to understand future advancements in hypersonic flight propulsion systems with the use of scramjet inlets is the ultimate aim. The focus of the work is narrowing the inlet flow up to the expected discrimination level and increasing the total pressure recovery.

as analysed by Shajahan S., et al. [84]. Nguyen P., et al.[85], investigates ramp-induced shock impingement effects in a strut-based/cavity-assisted scramjet combustor, optimizing combustion efficiency through specific slope angle configurations. The study of Samadhiya R. K., et al. [86] was to define enhancing combustion and mixing characteristics in Scramjet combustors through strut & wall injection techniques for better performance at specific outflow condition such as pressure, temperature and Mach number 2. Through calculations and experiments, M.R. Gruber et al. [87] investigated the cavity-based flame holder in a Mach 3 supersonic combustor. According to their research, cavities with shallower ramp angles had shorter residence times and higher drag coefficients. The effect of pylon mounted cavity in the combustor creates stream-wise vortices and augments the flame holder performance as numerically studied by S. Nayal et al. [88] and Hao Ouyang et al [89]. A numerical study comparing various geometrical cavity designs revealed that the combustion performance is significantly improved in combustors utilizing a cavity-strut configuration. Such enhancement is attributed to better fuel-air mixing and flame stabilization brought about by the combined effects of the cavity and strut design is reported by O. Reddy et al. [90] & S. Lakka et al. [91]. Sunwen Du et al. [92] carried out an exhaustive study on hydrogen jet mixing in a cavity flame holder with four different configurations. The study analyzed the flow and mixing zones within the cavity region and examined the impact of multiple jets on mixing performance.

According to Zhang et al.[93], the application of self-designed water plasma ignitors has proven effective in reducing ignition delay times and increasing combustion efficiency as well. In addition, the potential to greatly improve mixing efficiency (an important aspect of achieving effective combustion) by appropriate fuel injection strategies has been highlighted in studies on strut injectors, as demonstrated and quantified by Quan et al., [94]. By altering the inlet area and air intake height, Linqing, Q et al. [95] have numerically examined the performance of a novel strut-based scramjet combustor configuration and found a significant impact on combustion efficiency. In order to promote superior combustion performance using supersonic air flow to achieve High-Speed capability (HSc), V. Venkateshwaran et al. [96] conducted an experimental study of flame stabilization and shockwave mitigation by varying strut injection angles and the influence of injector design and angles in scramjet combustors. Scramjet combustors with cavity are complex systems, as the combustion processes in these geometries have a broad spectrum of behavior's influenced by inflow stagnation pressures that manifest differently on the flame structure and its intensity, and also oscillation behaviours during operation as analysed by Tai-Yu et al. [97]. T. Vanyai et al.[98] performed experimental tests for the combustion of ethylene in a semi-free jet kind of testing setup for axisymmetric scramjet combustors, thus imitating the conditions of a Mach 7-8 flight.

Kumaran K. et al. [99,117-121] have done a thorough study on the creation of hypersonic vehicles, primarily directed toward the requirement of efficient propulsion systems. In their research, they have pointed out the necessity for the quick and complete mixing and burning of one of the following: hydrogen, kerosene, or ethylene in a supersonic crossflow, while simultaneously employing different chemical kinetics and turbulence models.

Namrata B. et al. [100] studied how the combination of hydrogen and ethylene can be used as fuels for scramjets and concluded that the use of ethylene caused the recirculation zones to be more intense, whereas hydrogen gave rise to higher maximum pressure. The development of a new strut design and the impact of different backward-facing step heights and lengths on two-strut scramjet performance were investigated by Chunlei et al. [101].

Yu Jiang et al. [102] and Yeping Peng et al. [103] investigate the impact of upstream struts with multiple fuel injections inside supersonic combustion. The performance effect of a ramp location that is proximate to the strut in a non-reacting flow enhances the performance of the standard DLR model is studied by A. Antony A. et al. [104].

The study on a dual-strut with different angles and effect of wall injection is analysed numerically by G. Choubey et al. [105-106, 112] who observed that the dual-strut with zero degrees' deflection was having better performance characteristics. Then the implication effect of a diamond-shaped dual-strut is analysed by L. Sunnetha et al. [107]. G. Choubey et al. [108] describe computational simulations of a Mach 2.5 scramjet combustor with three central lobed struts, and the findings reveal that the three-strut injector beats the single-strut injector in terms of combustion efficiency. Another study describes how jet designs and total pressure ratio affect the strut's mixing characteristics, demonstrating that the vast subsonic area behind the strut increases fuel mixing and penetration as analysed by Yu Jiang et al. & Liang-ze Lu et al. [109, 123].

The investigation of varying inlet conditions to supersonic combustion of hydrogen fuel is studied by V. Hariharan et al. [110]. According to numerical calculations by Barzegar G. et al. [111], when the incident shock intersects the lower surface at a low angle of 15°, the maximum concentration of the hydrogen jet downstream drops dramatically, by as much as 30%.

Roga et al. & Obula Reddy K et al. [112, 116] found that the improvements in the fuel-air mixing and reduced ignition delays are key to efficient combustion. While vorticity and shock waves facilitate mixing, boundary cooling enhances scramjet performance. Verma et al. & V. Sharma et al. [113, 131-132] analyzed the performance of scramjet combustion with fuel injection. Air angle of attack effects are validated by computational approaches. Fuel schemes, isolators, and shockwave behavior are crucial. At -5°, maximum combustion efficiency is 93%, while at the same angle; mixing efficiency is 4% higher. Scramjet combustors utilize supersonic airflow for combustion, with regenerative cooling enhancing performance by cooling the combustor walls and improving mixing efficiency and chemical kinetics through high-temperature fuel injection as studied by Duo, Zhang et al. [114]. S.T. Surzhikov et al. [115]. validated the combustion process in a scramjet engine using a quasi-one-dimensional model and a two-dimensional model test using the Navier-Stokes equation. They also constructed a hypothetical model of the X-51 and assessed the results.

The combination of fuel injection techniques and cavity-based flame holders has significant attention for achieving efficient fuel injection and effective flame stabilization. The idea of cavity flame holders has been broadly dissected in scholarly articles [124–130] across various operating conditions to determine their impact on the flow field, ignition characteristics, fuel–air mixing, and overall combustion efficiency in scramjet combustors. Generally, a cavity flame holder entails a portion of the combustor wall withdrawn to form a hollow, which aims to produce a recirculation zone that not only traps the hot gases but also heats the fresh gases ensuring stable combustion in high-velocity flows. The flow thus recirculated is a flame stabilization technique, which finds application in scramjet engines, the air-fuel mix's residence time being very short due to the supersonic nature of the incoming air.

Scientists have experimented with different cavity shapes like shallow, deep, and tandem cavities to comprehend their effects on flow separation, shear layer formation, and pressure distribution inside the combustor. These factors, in turn, determine the ignition reliability and combustion stability as the hot gases trapped in the cavity improve the flame holding capability even in the case of hostile flow conditions.

The residence time of the supersonic airflow particularly is a factor that can tip the scale towards complete combustion as it is the one that controls the fuel-air mixture's time of stay in the high-temperature recirculation zone. By merely tuning the residence time coupled with the correct fuel injection orientation and cavity location, the combustion efficiency, thrust production, and thermal management in scramjet engines can be elevated to an impressive level.

Briefly, cavity flame holders represent a reliable passive combustion stabilization technique, thus, are instrumental in releasing the scramjet's limited combustor length's energy efficiently. The work in this area does not come to an end, they are continuously trying to figure out the best cavity designs and pairing them with cutting-edge fuel injection methods to attain stable, high-efficiency combustion at varying Mach numbers ranges. Results show that the adoption of cavity flame holders in scramjet combustors resulted in increased residence time in smaller length, improving overall performance. The adoption of cavity-flame holding in scramjet combustors essentially enhanced their work significantly by providing higher residence time for supersonic flow. Long stay time enables a larger extent of interaction between the oxidizer as well as the fuel, maximized ignition needs, together with enhanced combustion efficiency. Overcoming significant technologies of the future in supersonic combustion, cavity flame holding strives to advance the design as well as the potential within scramjet engine configurations that can be incorporated within high-speed propulsion systems.

Scramjets have represented the future promise of hypersonic flight with efficient propulsion off the atmosphere based on oxygen. However, efficient combustion and injection of the fuel, however, presents a significant issue with highly extreme flow based on supersonic flow. Specifically, very limited residence time with complex interaction with the shock as well as turbulence makes the sustaining of efficient combustion complex.

Table 2.3: Comparative Summary of Scramjet Combustion Research

Configuration / Technique	**Research Focus / Objective**	**Key Findings / Outcomes**	**References**
Strut-Based Combustors	Effect of strut location, ramp angles, and fuel injection position	Strut injectors enhance fuel–air mixing via induced shear layers; optimized injection angles improve combustion	[71, 75–82, 85–87, 90–91, 94–96, 102–109,]

2.8 Overall Objectives

- Performance enhancement investigation of a Scramjet Inlet
- Numerical investigation of a Scramjet Combustor
- Performance assessment of a Scramjet Engine by integration of individually adopted high performing Scramjet Inlet and Combustor using numerical investigation.

CHAPTER 3

SIMULATION METHODOLOGY

3.1 Introduction to Computational Fluid Dynamics (CFD)

Fluid dynamics, along with its mathematical interpretation, computational fluid dynamics (CFD), deals with the change in fluid behavior over time, which is described by partial differential equations. By using mathematical modelling, numerical methods, and software tools, CFD is able to penetrate the flowing dynamics and deliver both qualitative and quantitative predictions of fluid flows.

The CFD analysis process in FLUENT software encompasses the steps of conceptualizing a model, meshing it, defining boundary and solver conditions, solving for variables, visualizing results, and storing data. This is a simulation process for fluid flow behavior of different kinds of problems, which is directed by the governing equations in FLUENT.

3.2 Governing Equations in CFD

In computational fluid dynamics, we typically solve three equations: the Continuity equation, the Momentum equation (also known as the Navier-Stokes equation), and the Energy equation. The numerical simulations in the present work are carried out using the commercial CFD solver ANSYS Fluent [11,12], which is extensively employed in both academic research and industrial applications for modelling complex flow phenomena. The solver numerically integrates the fundamental conservation equations governing fluid motion. These include the continuity equation for mass conservation and the momentum equations for all simulated flow cases. In situations involving compressibility or thermal effects, the energy conservation equation is additionally solved. Furthermore, for reacting or multi-species flows, transport equations for individual chemical species are incorporated. When turbulence effects are present, additional transport equations associated with the selected turbulence model are also resolved.

3.2.1 Governing Equations

The continuity equation (1) represents the general form of mass conservation and is applicable to both incompressible and compressible flows. Source terms may be included

$$\Omega_D = \frac{1}{(T_D^*)^{0.145}} + \frac{1}{(T_D^*+0.5)^2} \tag{22}$$

where T_D^* is the dimensionless temperature given by,

$$T_D^* = \frac{T}{\left(\frac{\varepsilon}{K_B}\right)_{ij}} \tag{23}$$

K_B is the Boltzmann constant, which is defined as the gas constant, R, divided by Avogadro's number. $\left(\frac{\varepsilon}{K_B}\right)_{ij}$ in Kelvin for the mixture is the geometric average:

$$\left(\frac{\varepsilon}{K_B}\right)_{ij} = \sqrt{\left(\frac{\varepsilon}{K_B}\right)_i + \left(\frac{\varepsilon}{K_B}\right)_j} \tag{24}$$

3.4.2 Turbulence Chemistry coupling

When turbulence–chemistry interaction (TCI) modelling is not explicitly employed, chemical source terms are evaluated using finite-rate reaction kinetics without accounting for turbulent fluctuations. This approach is suitable for cases where chemical time scales are comparable to or slower than turbulence time scales, or when detailed chemical mechanisms are involved.

The net production rate of each species is obtained by summing its contributions across all participating reactions, with porosity effects included where applicable. For reactions in a porous medium, the net source due to reactions is obtained by multiplying the reaction rates with porosity:

$$R_i = M_{w,i} \sum_{N_R}^{r=1} \varepsilon \widehat{R_{i,r}} \tag{25}$$

where $M_{w,i}$ is the molecular weight of species i, ε is the porosity, and $\widehat{R_{i,r}}$ is the molar rate of creation/destruction of species i in reaction r. Reaction may occur in the continuous phase at wall surfaces.

Consider the rth reaction written in general form as follows:

$$\sum_{i=1}^{N} v'_{i,r}\mathcal{M}_i \rightleftharpoons \sum_{i=1}^{N} v''_{i,r}\mathcal{M}_i \tag{26}$$

where, N is the number of chemical species in the system, $v'_{i,r}$ is the stoichiometric coefficient for reactant i in reaction r, $v''_{i,r}$ is the stoichiometric coefficient for product i in reaction r and $\mathcal{M}_i$ is the symbol denoting species.

Chemical reaction in equation (26) is described using general stoichiometric formulations, applicable to both reversible and irreversible reactions. For irreversible reactions, the backward rate constant, $k_{b,r}$, is zero. All chemical species in the system are included in the summations in Equation (26), but only those that show up as reactants or products will have stoichiometric coefficients that are not zero. The molar rate of creation/destruction of species i in reaction r is given by,

$$\widehat{R_{i,r}} = \Gamma\left(v''_{i,r} - v'_{i,r}\right)\left[k_{f,r}\prod_{j=1}^{N}\left[C_{j,r}\right]^{\eta'_{j,r}} - k_{b,r}\prod_{j=1}^{N}\left[C_{j,r}\right]^{\eta''_{j,r}}\right] \tag{27}$$

where, $C_{j,r}$ is the molar concentration of species i in reaction r (kmol/m^3), $\eta'_{j,r}$ is the rate exponent for reactant species j in reaction r, $\eta''_{j,r}$ is the rate exponent for product species j in reaction r and $k_{f,r}$ & $k_{b,r}$ are the forward and backward rate constant for reaction r. Observe that the rate exponent for the reverse reaction component in Equation (27) corresponds to the stoichiometric coefficient of the product species ($v''_{j,r}$).

Γ represents the net effect of third bodies on the reaction rate. This term is given by

$$\Gamma = \sum_{j}^{N} \gamma_{j,r} C_j \tag{28}$$

where $\gamma_{j,r}$ is the third-body efficiency of the jth species in the rth reaction. Third-body effects are by default excluded from the solver's computation of response rate. However, if you have statistics for third-body efficiencies, you can choose to incorporate their impact. The forward rate constant for reaction r, $k_{f,r}$, is computed using the Arrhenius expression:

$$k_{f,r} = A_r T^{\beta_r} e^{\left(-E_r/RT\right)} \tag{29}$$

where, A_r is the pre-exponential factor (consistent units), β_r is the temperature exponent (dimensionless), E_r is the activation energy for the reaction (J/kmol), R is the universal gas constant (J/kmol-K). The backward rate constant for reaction r, $k_{b,r}$, is automatically calculated from the forward rate constant using the following formula if the reaction is

	and streamline curvature. Includes an additional term for turbulent viscosity correction.	separated and recirculating flows• Accounts for smaller-scale turbulence effects	• Still assumes isotropy in turbulence	layer flows • Moderate swirl combustor flows
Realizable k–ε	Incorporates a new formulation for turbulent viscosity and an alternative ε equation derived from the mean-square vorticity transport equation.	• More accurate in predicting flow separation and rotation-dominated flows • Provides realizable constraints (physically consistent turbulent stresses) • Better convergence and numerical stability compared to RNG	• Cannot be used with multiple rotating reference frames • Limited accuracy in transitional or wall-bounded low-Re regions	• Aerospace intakes and exhausts • Combustion chambers with moderate swirl • External aerodynamics where separation prediction is important
Standard k–ω	Uses specific dissipation rate (ω) instead of ε, offering better near-wall treatment.	• Superior prediction of wall-bounded and low-Reynolds-number flows	• Sensitive to free-stream ω values (requires careful initialization) • May	•Turbomachinery flows • Boundary-layer flows under adverse pressure

		• Captures near-wall gradients effectively • Handles adverse pressure gradients better than k–ε	overpredict turbulence away from walls	gradients • Transition and separation onset studies
SST (Shear Stress Transport) k–ω	Combines the advantages of k–ε and k–ω models using a blending function; k–ω near the wall and k–ε in the free stream.	• High accuracy for boundary layers under adverse pressure gradients and separation • Smooth transition between near-wall and free-stream regions • Well-suited for external aerodynamics and compressible flows	• Increased computational cost• Dependent on near-wall mesh resolution • Less effective in pure free shear flows	• External aerodynamics (airfoils, vehicles) • Supersonic and hypersonic internal flows (scramjets) •Turbomachinery and propulsion systems

In CFD, 'k' represents turbulence kinetic energy, signifying the variance in velocity fluctuations, typically measured in m^2/s^2. Meanwhile, 'ε' denotes turbulence eddy dissipation, indicating the rate at which velocity fluctuations dissipate, measured in m^2/s^3. The turbulent kinetic energy equation simplifies several terms for modeling purposes. It includes production of 'k', dissipation (ε), and terms describing 'k' transport via molecular and turbulent diffusion. Most turbulence models solve this equation, requiring calculation of 'ε' and eddy viscosity ('ut'). The standard k-ε model in FLUENT is commonly used for high Reynolds number flows, where turbulence is nearly isotropic and the energy cascade proceeds in local equilibrium. FLUENT also offers the RNG model.

3.5 The Advantages of CFD

Briefly, In real life, there are three good reasons for believing in CFD:

Insight: CFD gives special insight into complex systems, revealing phenomena which are difficult to observe otherwise. This additional knowledge provides helpful hints to designs, allowing improved visualization and comprehension.

Foresight: Being a powerful predictive tool, Computational Fluid Dynamics (CFD) enables the engineer to solve and analyze complex fluid flow and thermal phenomena before any physical model are built. CFD offers the ability to perform real-time "what-if" analysis, where parameters of design—geometry, boundary conditions, properties of the material, and operating conditions—can be modified in a systematic way to predict how they will affect overall system performance. This forecasting ability contributes to a better understanding of fluid flow, combustion, and aerodynamic performance under real conditions. In fact, the designer finds potential deficiencies of the design by the help of various flow simulations, such as zones of high-pressure, shock, separation, or thermal hotspots, which can be corrected at an early stage.

Additionally, CFD allows for parametric and optimization studies with the potential of virtual experiments of hundreds of design variations within a very short time. This makes decision-making process noticeably easier as an engineer gets a chance to evaluate different design options and choose the one that ensures a good balance of performance, stability, and efficiency. Predictive insight turns out to be extremely important in high-speed propulsion systems like scramjets where the experimental approach is limited by the factors of cost, safety, and the availability of hypersonic facilities.

To a large extent, CFD can be considered as a digital laboratory that makes it possible to give very close to reality predictions of aerodynamic, thermal and chemical phenomena which, in turn, govern design improvements and guarantee strong system performance even before fabrication or testing.

Efficiency: The use of CFD in the design cycle leads to a significant reduction of engineering time and expenses besides improving the general effectiveness of the process. Experimental testing based on physical models usually requires the production of numerous mock-ups and the execution of large-scale combustion or wind tunnel tests that are not only costly but also take up a lot of time. As far as these methods are concerned, CFD through fast iterations and virtual testing provides deep insight at a fraction of the cost of physical models.

The acceleration of the design–analysis–optimization cycle leads to lower product delivery times and, as a result, faster improvements through the design–analysis–optimization cycle. Pressure distribution, flow fields, temperature distribution, as well as other standard parameters can be readily accessed by designers, allowing them to pinpoint performance problems early in the development process. This leads to an overall improvement in design quality over time while at the same time avoiding costly reworks downstream.

CFD software can be linked with the present Computer-Aided Design (CAD) software and allow optimization algorithms to be integrated for the automatic and parallel simulation of large numbers of design configurations. The end result of this is a faster approach to optimal designs with better performance and lower material and production costs.

At the end of the day, CFD elevates the project stages in a productive, adaptable, and scalable manner enabling the execution of detailed fluid flow analyses. It keeps system design going, lessens the time needed for changes or upgrades, and, most importantly, the market launching process is accelerated, which is a crucial advantage in the engineering sectors that are highly competitive, like aerospace propulsion, automotive styling, and energy systems.

3.6 Applications of CFD

Computational Fluid Dynamics (CFD) is one of the most potent analysis techniques for understanding and anticipating fluid behaviors in a variety of engineering situations. When its main governing equations of fluid motion, Navier–Stokes equations, and energy and species transport equations are solved numerically—CFD can give a full quantitative picture of the flow fields along with the heat and mass transport, chemical reactions, and phase interactions. Through the application of this numerical method, engineers are able

CHAPTER 4
NUMERICAL SIMULATION OF SCRAMJET INLET

4.1 Introduction

By definition, scramjets constitute the best form of air-breathing hypersonic propulsion and, as a result, are considered the most likely source of the vital fuel combustion mechanism in question, at speeds greater than Mach 5. One of the main differences between scramjets and other types of jets is that the former does not employ rotating compressors for squeezing the incoming air. What scramjets do is utilize the incredibly energetic high-speed airflow, which in turn performs the compression.

At these speeds, the air, which is supplied to the engine, is extremely fast and as a result, it is accompanied by very high pressures and temperatures. Therefore, the inlet is constructed with a great deal of accuracy so that the air, which is trapped and compressed, is also brought to the burner in a controlled and efficient manner. In case of inefficiency, poor combustion, reduced thrust, or engine unsteadiness could be possible consequences.

The inlet represents the connection layer between the scramjet engine and the rest of the propulsion system, and its performance is the determinant of the whole system's efficiency. An appropriately functioning inlet will ensure the sufficiently slowing down and compression of the air with a minimum energy loss. The compression being done in such a cheap way is what finally makes it possible for scramjets to offer the large thrust required for hypersonic flight at steady states.

The construction of such an inlet is a challenge far beyond simple. The engineers need to be extremely careful concerning many aerodynamic, thermodynamic, and material issues. The occurrence of shock waves, boundary layers, and heat under these extreme conditions can, without a doubt, have very negative effects on performance. Small errors in design can lead to flow separation or unstart of the inlet occurrence with a huge negative effect on the stability of the engine [67-70].

To tackle such problems, the design of modern scramjet inlets largely relies on sophisticated Computational Fluid Dynamics (CFD) modeling and ground wind tunnel testing. CFD lets engineers visualize the airflow in the inlet for various flight conditions without the high cost of building a physical model for each condition, while the experimental validation serves as a bridge to practical implications. By alternately using both methods, engineers are able to modify inlet shape for optimal compromise between compression efficiency, shock management, and heat shielding.

Its form is a trade-off between aerodynamic accuracy, heat resistance, and structural integrity, and it is one of the main components in the highly-efficient hypersonic flight quest.

4.2 Design Methodology

This scramjet research includes the use of ANSYS Fluent for design, analysis, and modeling of the scramjet combustor and inlet through a CFD-based systematic approach. The method starts off with the determination of engineering targets, flow regime, operating conditions, and significant performance measures ones like pressure recovery, mixing efficiency, and thrust. Subsequently, the geometry is either created or imported with all the essential details of the critical flows i.e. compression ramps and cavities accurately depicted. The figure is then subdivided into a grid with refinements in the locations of expected shock interactions and boundary layers.

After that, the properties are assigned and authentic boundary conditions are applied to simulate real operating conditions. The work is done by selecting the proper solvers and physical models, while convergence and stability are constantly being checked. Post-processing mainly revolves around the visualization of the flow-fields, extraction of the performance metrics, and the examination of the flow structures like shock waves and recirculation zones. By analyzing these data, design iterations are decided upon to further optimize the performance. Such decisions may involve changes to ramp angles, adjustments to the cavity geometries, or bleed gap manipulations. Ultimately, the simulation results are verified against the experimental or benchmark data to confirm the precision and reliability of numerical predictions.

applied to the left, upper, and lower boundaries, while the right-side boundary utilized pressure outlet conditions.

Numerical simulations were conducted with the use of ANSYS Fluent R18 using density-based calculations with the Shear Stress Turbulence k-ω two-equation model. The fluid was modeled as an ideal gas with the viscosity based on the three-coefficient Sutherland's method.

ANSYS Fluent is a highly regarded computational fluid dynamics package that generates accurate, detailed simulations of the complex flow dynamics associated with scramjet engine flow. By employing the k-ω turbulence model and Sutherland's method of viscosity, the study assured a robust and reliable modeling of the fluid behavior in extreme conditions, a pivotal precursor for optimizing the intake and combustor designs.

Set-up parameters:

The set-up parameters used for the validation of the inlet is shown below.

Table 4.1 Set-up parameters

Sl. No.	General parameters	
1.	Solver	Density-based
2.	Time	Steady
3.	Energy Equations	On
4.	Model	SST K-omega
5.	Wall function	SST
Spatial discretization		
6.	Gradient	Green gauss cell based
7.	Pressure	Second order
8.	Momentum	Second order upwind
9.	Temperature	Second order upwind
10.	Density	Second order upwind
Method solution		
11.	Formulation	Explicit
Initialization		
12.	Standard initialization from left farfield	

Mach Contours:

CFD simulations displaying shock wave static pressure, temperature, and Mach number contours about the inlet geometry provide a detailed visualization to evaluate the inlet's ability to maintain supersonic flow in the scramjet engine is shown in Figure 4.4. Comparing the simulation data of the relevant parameters with the ideal Mach number will illuminate flow inefficiencies and areas for inlet design improvement. These data will serve for implementing targeted design changes for optimized flow conditioning. After the new inlet is designed, new CFD simulations are conducted to validate the increased performance. This iterative process ensures that the inlet is fine-tuned to better the overall scramjet engine performance. The ultimate goal during this project is to help develop and iterate an optimized scramjet engine design.

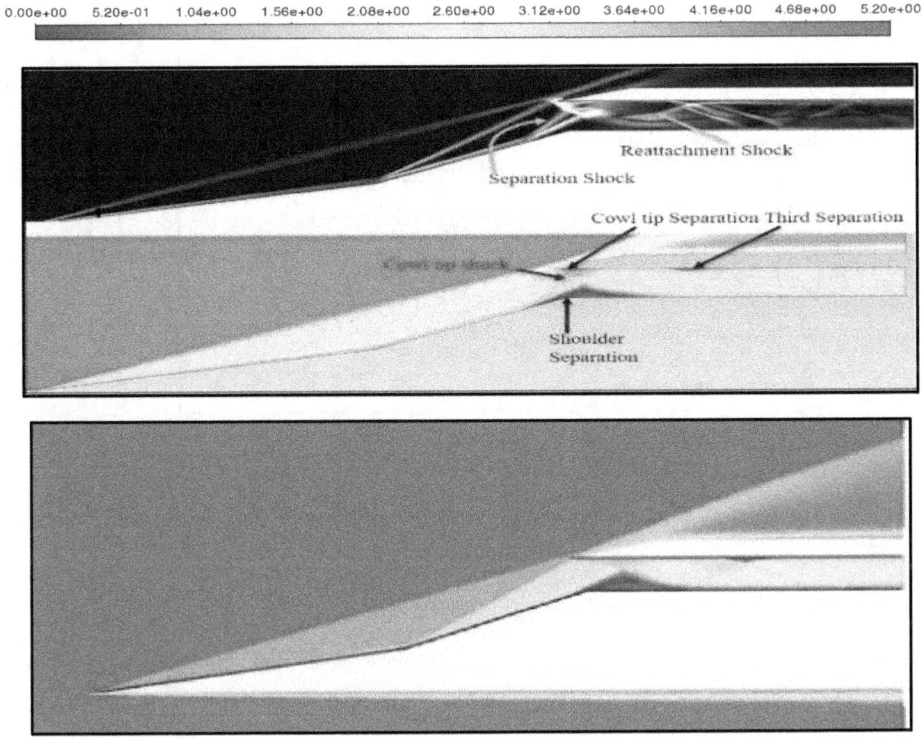

Fig 4.4: Comparison of Mach contours of Validation Paper and present simulation

Validation Plot:

Fig. 4.5 shows the static pressure variation along the ramp surface in flow direction, which corresponds to Azam Idris et al. [31] computations and experiments. The correlation between observations and earlier studies results in a good comparison, thus confirming the validity of the solver employed in this study. Observed pressure change confirms that CFD simulated flow properly represents the flow behavior inside the scramjet engine. Validation is essential for confirmation of modeled result accuracy and confidence in the model's prediction potential.

Using an accurate model leads to an improved inlet design by being able to improve location of inefficiencies. Once the inlet geometry was improved from validated simulation then the performance and efficiencies can be improved greatly with a scramjet engine. This serves to continually improve scramjet technologies into better designed and more reliable engines.

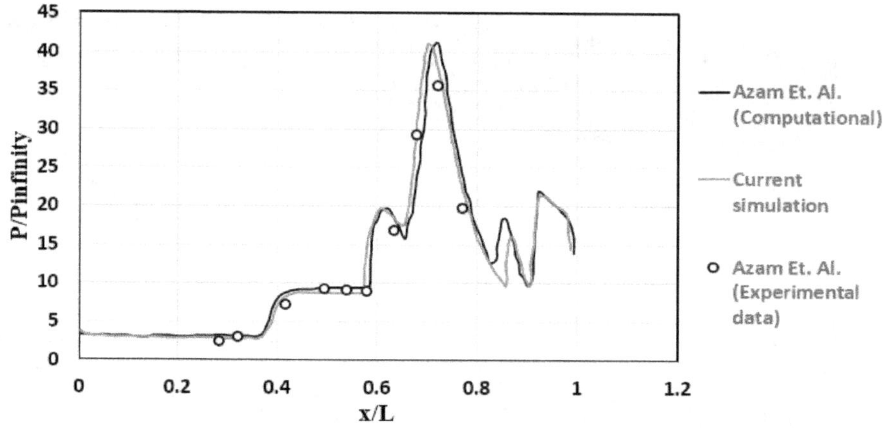

Fig 4.5: Non-dimensionalized pressure distribution on ramp surface for validation

4.4 Grid Independence Test

To make sure that the mesh size had no effect on the simulation results, the grid independence test was run. This was done by employing varying numbers of elements, including coarse (144,436 elements), medium (224,189 elements), fine (308,585 elements), and very fine (468,377 elements) meshes as shown below in Figure 4.6. The geometry and setup of the simulations remained consistent across all tests, with the only variable being the number of mesh elements, as illustrated in Figure 4.1.

The goal of the study was to find the point at which grid independence was achieved by methodically increasing the mesh density until additional refinement no longer significantly changed the results. Following the completion of the grid independence test, it was determined that the mesh with 308,585 elements produced results that closely matched both the computational and experimental data. This level of mesh refinement provided a good balance between computational efficiency and accuracy.

Additionally, to ensure the accuracy of the simulations, a wall Y+ value below 1 was maintained in all computations. This low Y+ value is critical in capturing the near-wall flow characteristics accurately, which is especially important in high-speed flow simulations typical of scramjet engines. Maintaining a Y+ value below 1 ensures that the turbulence model used in the simulations can accurately resolve the boundary layer effects, contributing to the overall reliability of the simulation results.

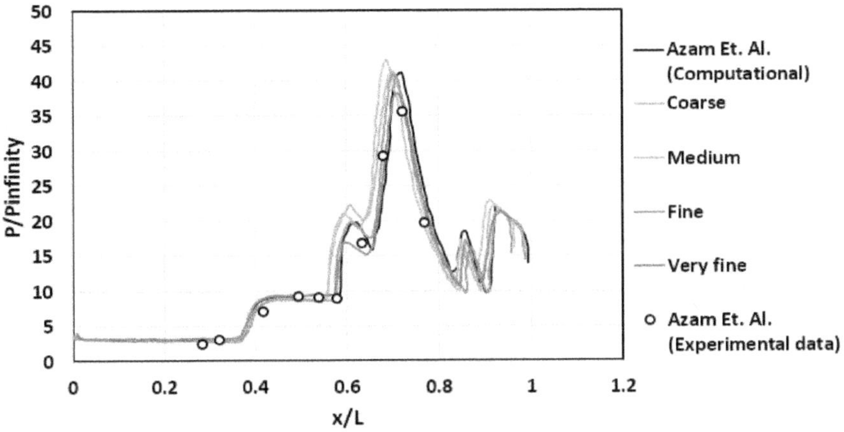

Fig 4.6: Non-dimensionalized pressure distribution on ramp surface for GIT plots

Table 4.2: Table showing the Y+ value for different mesh configurations

Case	Number of elements	Y+ value
Coarse	144436	Near 1
Medium	224189	1
Fine	308585	Below 1
Very fine	468377	Below 0.5

4.5 Different Inlet Performance Enhancement Techniques

Improvement of passive inlet performance methods covers a great number of techniques and solutions, which basically aim at enhancing the aerodynamic performance and energy efficiency of air intakes without any need of adding active control mechanisms. Following is the overview of the methods considered:

Cowl Deflection Angle:

Definition: The cowl deflection angle is the angle at which the cowl, the movable part of the air intake system, is deflected.

Purpose: By changing the cowl angle the airflow can be regulated inside the duct thus shock waves can be weakened, pressure recovery can be optimized and aerodynamic losses can be decreased.

Application: It is highly effective in supersonic engines like scramjets where it is absolutely necessary that shock waves be controlled to get high engine efficiency.

Passive Bleed

Definition: Passive bleed is the concept of a multitude of small holes or vents that are distributed over the inlet surface are meant to bleed some air from the surface boundary layer.

Purpose: By changing the mass flow through the leak different effects can be mitigated, e.g. separation bubbles can be weakened and shock waves can get less intense.

Application: The technology is very popular for use in supersonic and hypersonic inlets, where it is possible to increase compression and achieve higher pressure recovery ratios.

Cavity:

Definition: A cavity is a geometrically defined pocket or recess, mostly used in an inlet.

Purpose: Having changed the flow patterns the device can reduce the generation of shock waves and increase pressure recovery.

Application: Basically, cavities are widely applied in the design of the inlet geometrical shapes to increase aerodynamic performance.

Cowl Shape (3D):

Definition: The three-dimensional cowl shape goes beyond the cowl deflection and cowl geometry, describing the 3D form of the cowl including the location and orientation of the parts.

Purpose: By increasing shock control and creating new airflow patterns cowl shape can be used to achieve compression and pressure recovery.

Application: The technology is used in the most advanced inlet conceptualization stages to achieve the best aerodynamic performance in diverse weather and flight conditions.

Ramp Deflection Angle:

Definition: Similar to the cowl deflection the term 'ramp deflection' refers to the change of the angle of a ramp, a device that redirects and compresses the incoming air, with regard to the air flow.

Purpose: Shock wave shift and changes in air stream due to ramp deflection angle help attain an efficient and justifiable compression of the needed air volume.

Application: The design of sophisticated supersonic and hypersonic inlets extensively utilizes the science of the compression process management via the use of the ramp deflection angle.

Vortex Generators:

Definition: Small aerodynamic elements precisely arranging in order to achieve the creation of a definite vortex in the airflow.

Purpose: The transition from laminar to turbulent boundary layer can be mastered and stabilized with the help of VG's leading to significant separation reduction and overall improvement of aerodynamic performance.

Application: VG's installation is not limited to one field but on the whole the aviation industry, i.e. in the wings, empennage and even in the inlets as well, they serve the purpose of airflow control and efficiency enhancement.

By solving most of the aerodynamic problems, these passive techniques extend the possibilities of air intake optimization and lead to greater overall performance of high-speed propulsion systems.

passive bleed is shown below.

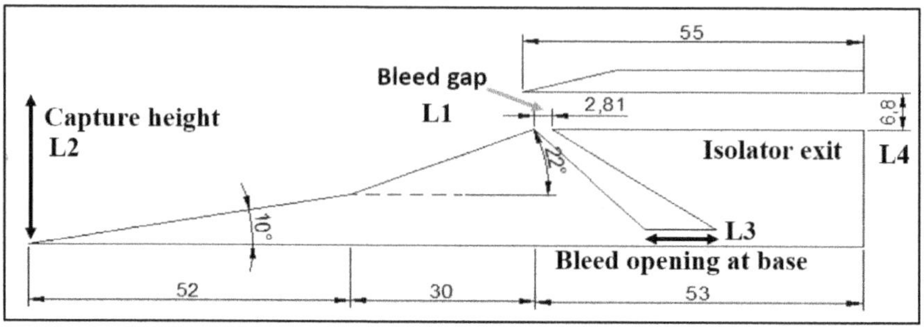

Fig 4.7: Bleed geometry dimensions

Table 4.4 Bleed dimensions

Length	Description
L1	Bleed gap
L2	Capture height
L3	Bleed opening at base
L4	Outlet
Bleed gap applied in the current simulations (L1) for capture height 28.09mm with reference from literature[40,43]	
1.404 mm	
2.809 mm	
4.213 mm	
5.618 mm	

Mesh, Boundary Conditions and Set-up Parameters: The mesh, boundary conditions and set-up parameters used for the inlet with passive bleed simulations are same as that of the validation case.

Results:

Mach Contours:-

Mach contours show that adding the passive bleed significantly reduces the separation bubble caused by lip shock on the ramp and lessens the normal shock at the isolator section. This results in an enhanced pressure recovery at the inlet outlet, a conclusion further substantiated by accompanying plots shown below in Figure 4.8(a-e).

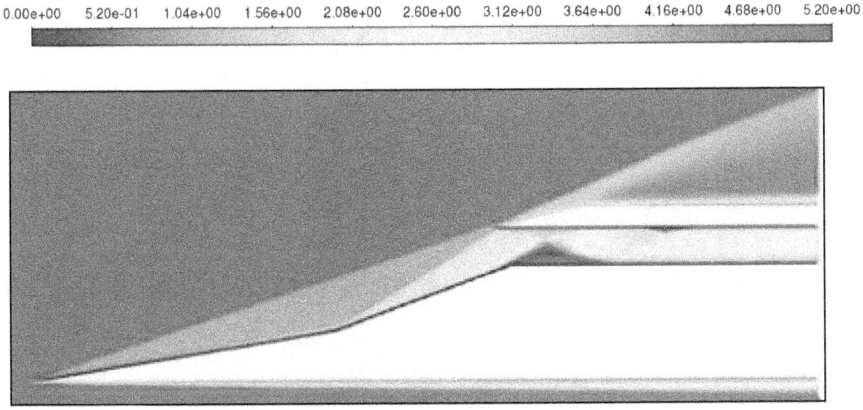

a) Mach contour base case

b) 1.404 mm bleed

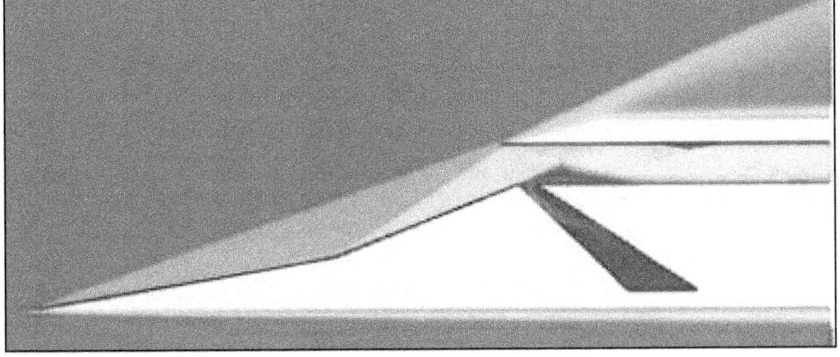

c) 2.809mm bleed

d) 4.213 mm bleed

e) 5.618mm bleed

Fig 4.8 Mach contours representing the change in flow physics for different bleed gaps as compared to base case

Plots:-

Plot of x/L vs Pstatic/Pinfinity at ramp interior region:

Typically, a specific number of shocks perform the compression in the scramjet inlet. The Shock Boundary Layer Interactions within the inlet-isolator body are depicted below in the Figure 4.9.

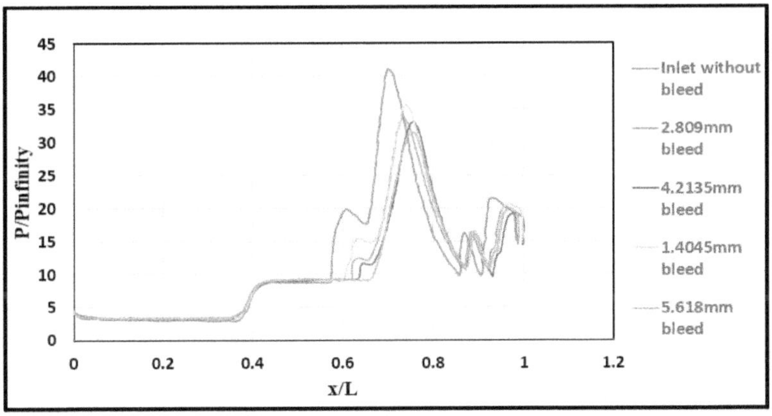

Fig 4.9 Non-dimensionalized pressure distribution on ramp surface for a) Different bleed gaps,

The plot depicts the variation of Pstatic/Pinfinity along the length (x/L) of the scramjet inlet. Specifically, the segment from position 0 to 0.4 corresponds to the Pstatic/Pinfinity at the first ramp (10-degree ramp), and the segment from 0.4 to 0.6 represents the Pstatic/Pinfinity at the second ramp (22-degree ramp). Beyond x/L 0.6, there is a noticeable surge in pressure, indicative of the Pstatic/Pinfinity at the isolator section. We can observe that there is a sudden increase in the Pstatic/Pinfinity at the isolator section which leads to pressure loss. The diminished efficiency of the scramjet inlet can also be attributed to three types of flow separation: shoulder separation, cowl tip separation, and the third separation occurring at the isolator. Our aim is the reduce this pressure loss and thereby increasing the pressure recovery on the inlet which bin turn increases the overall efficiency of the inlet which is obtained and is shown in the results.

Plot of Ptotal/Pstagnation vs Position at Outlet:

Plot of Ptotal/Pstagnation vs Position at Outlet is shown in Fig 4.10. The pressure recovery at the Scramjet Inlet's outlet is essentially depicted in this plot. The ratio of the stagnation pressure at the inlet's outlet to the initial pressure when the air first entered the inlet is known as the pressure recovery. It illustrates how the compression process results in a decrease in overall pressure.

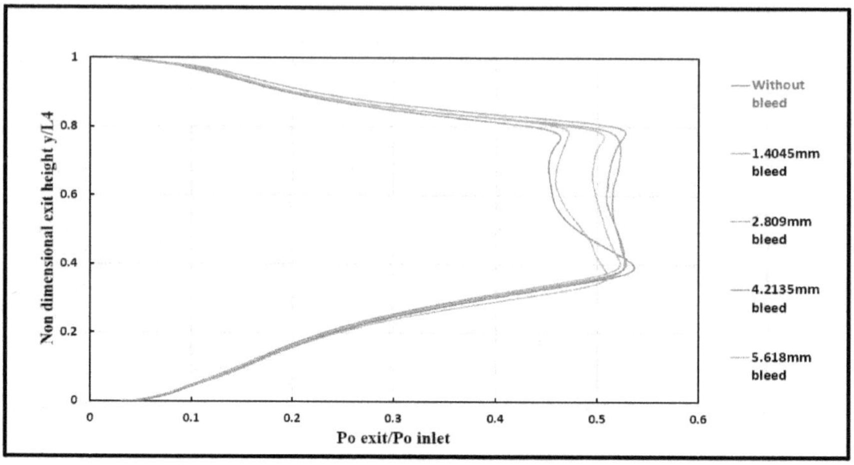

Fig 4.10 : Comparison of total pressure recovery for different bleed gaps

Table 4.5: Table shows the increase in pressure recovery for different bleed cases

Case	Pressure Recovery	Percentage Increase in Pressure Recovery
Validation Case	0.287144	NA
1.4045mm bleed	0.2966682	3.3168
2.809mm bleed	0.3465767	20.6978
4.2135mm bleed	0.3416698	18.989
5.618mm bleed	0.3288346	14.519

4.6.3 Cowl Deflection Angle

The cowl deflection geometry applied in the present study have been integrated into the geometry, as per the framework outlined by S, Das et al. [50]. Azam Che Idris et al.'s inlet geometry [31] lacked a cowl deflection angle. In this study, the authors have added five distinct cowl deflection angles: one, two, three, four, and five degrees, respectively. The following subsections then analyses and report on the impact of cowl deflection angles on the inlet performance.

Geometry: The cowl deflection dimensions used in our work is incorporated in the geometry with reference to S. Das[50]. The geometry on scramjet inlet with incorporation of cowl deflection angle is shown below in Figure 4.11.

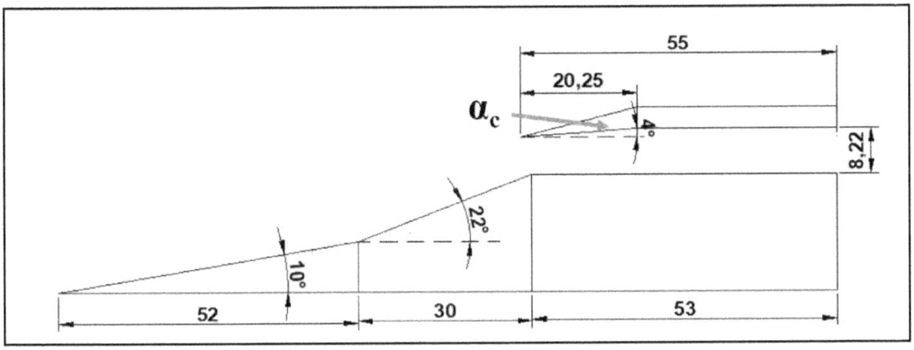

Fig 4.11: Cowl deflection angle geometry dimensions

Mesh, Boundary Conditions and Set-up Parameters: For the inlet with passive bleed simulations, the mesh, boundary conditions, and setup parameters are identical to those used in the validation case.

Mach Contours:

The main focus of the study is to evaluate the influence of various cowl deflection angles on scramjet inlet performance with a major emphasis on the Mach contours. Experimental results indicate that the use of a certain cowl deflection leads to an increase in pressure recovery at the isolator exit as a result of the reduction of lip-shock-induced separation on the inlet ramp and the lessening of shock effects at the isolator section. A series of plots strengthen this claim as shown below in Figure 4.12(a-e).

Comparison between the scenarios of cowl deflection angles omission and presence unveils prominent variations in the flow separation patterns that include shoulder separation, cowl tip separation, and isolator separation. The creation of a cowl deflection angle at the inlet throat area lessens the intensification of the shoulder separation and, consequently, the shock train behavior in the isolator section is affected.

Additional examination determines the effect of different cowl deflection angle settings on pressure recovery at the isolator exit that lead to performance optimization of the inlet.

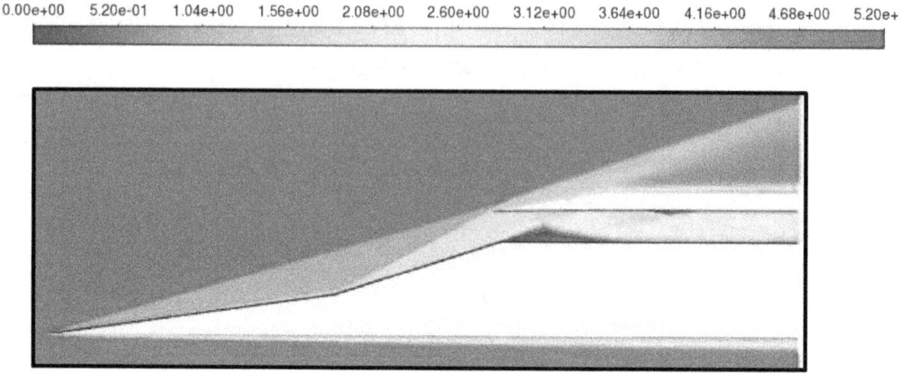

a) 1 degree

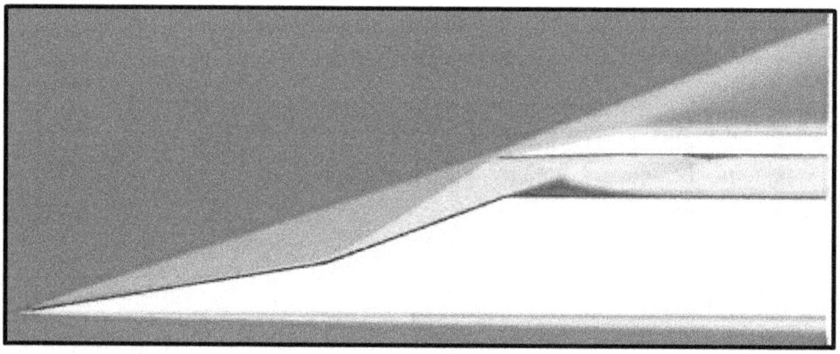

b) 2 degree

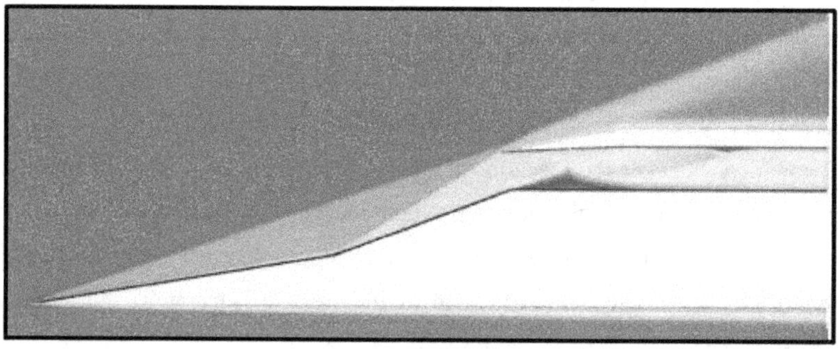

c) 3 degree

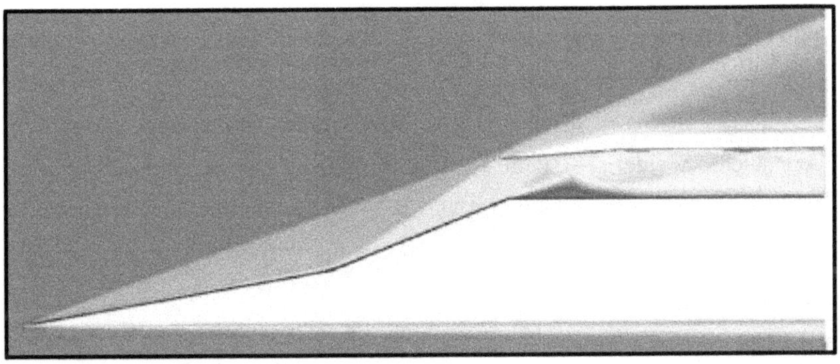

d) 4 degree

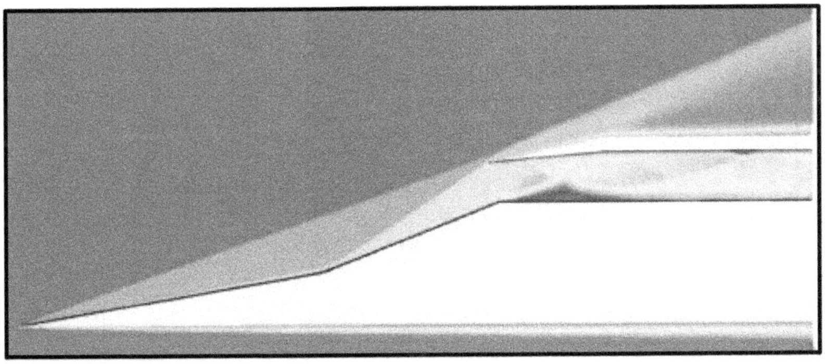

e) 5 degree

Fig 4.12: Mach contours representing the change in flow physics for different cowl deflection angles as compared to base case

Effect of cowl deflection angle on surface pressure distribution on ramp:--

Typically, a specific number of shocks perform the compression in the scramjet inlet. Figure 4.13 illustrates the Shock Boundary Layer Interactions within the inlet-isolator body by showing the non-dimensionalized pressure distribution on the ramp surface for different cowl deflection angle configurations. The surface pressure distribution variation along the scramjet inlet's length (x/L) is plotted. Specifically, the segment from position 0 to 0.4 corresponds to the surface pressure distribution at the first ramp (10-degree ramp), and the segment from 0.4 to 0.6 represents the surface pressure distribution at the second ramp (22-

degree ramp). Beyond x/L 0.6, there is a noticeable surge in pressure, indicative of the pressure distribution at the isolator section. We can observe that there is a sudden increase in the surface pressure distribution at the isolator section which leads to pressure loss. The diminished efficiency of the scramjet inlet can be attributed to three types of flow separation: shoulder separation, cowl tip separation, and the third separation occurring at the isolator. Our aim is the reduce this pressure loss and thereby increasing the pressure recovery on the inlet which in turn increases the overall efficiency of the inlet which is obtained and is shown in the results.

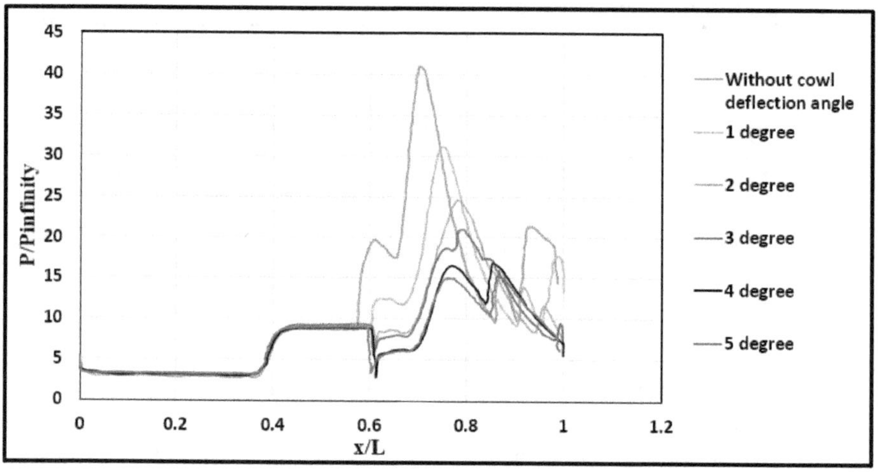

Fig 4.13 Non-dimensionalized pressure distribution on ramp surface for different cowl deflection angles.

Effect of different cowl deflection angles on the total pressure recovery at isolator exit:-

Figure 4.14 presents a comparison of total pressure recovery for the characterization of different configurations of cowl deflection angle in free flow at the exit. This figure illustrates the total pressure recovery at the Scramjet Inlet outlet, which is defined as total pressure recovery (π), the measured ratio of the stagnation pressure measured at the outlet of the inlet and the initial stagnation pressure when the air enters the inlet. Total pressure recovery is a measure of the total pressure loss during the process of compression. The y-axis is defined by the length L4 (refer Fig 4.6) where total pressure recovery is measured for the bleed and non-bleed cases, a crucial measure for assessing total pressure loss in the compression process. The interaction between the shockwave and the boundary layer can very easily disrupt total pressure efficiency that is a parameter of great importance in this work. Consequently, pressure recovery becomes a very important aspect in evaluating the pressure losses in the inlet- isolator system that is instrumental in upgrading the inlet's overall efficiency. Here are the pressure recovery figures for the validation case and the scenarios with the cowl deflection angles.

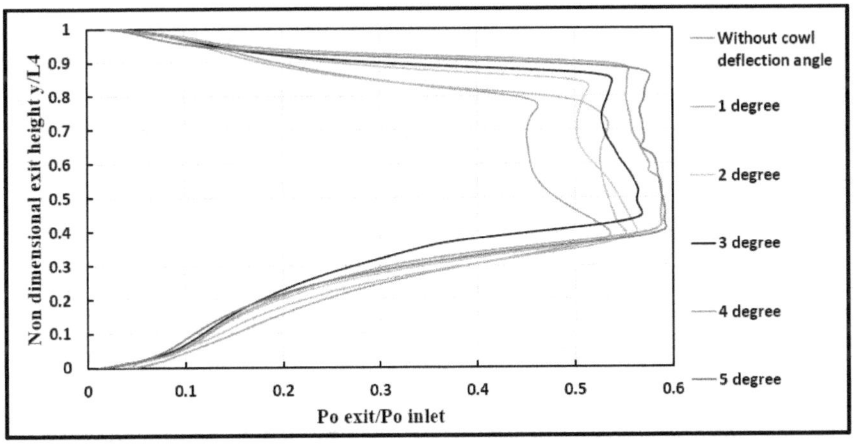

Fig 4.14 Comparison of total pressure recovery for different cowl deflection angles

Table 4.6: Percentage increase in total pressure recovery for different cases		
Case	Total pressure recovery Π	Percentage increase in total pressure recovery
Validation case	0.287144	NA
1 degree	0.302742	5.4321%
2 degree	0.348722	21.4449%
3 degree	0.308617	7.4781%
4 degree	0.379791	32.2649%
5 degree	0.34167	18.989%

4.6.4 Integrated bleed and cowl deflection angle configuration: This analysis involves incorporating the optimal bleed case, with a bleed size of 2.809mm, and the optimal cowl case, with a cowl deflection angle of 4 degrees.

Geometry

This analysis involves incorporating the optimal bleed case, with a bleed size of 2.809mm, and the optimal cowl case, with a cowl deflection angle of 4 degrees, as depicted below in Figure 4.15.

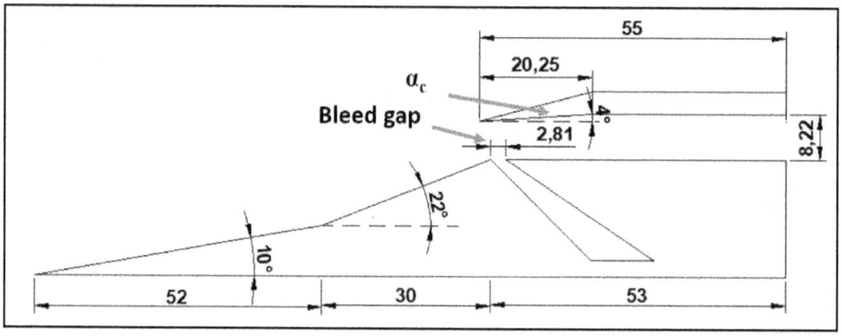

Fig 4.15: Best cases of Bleed(L1) and Cowl deflection angle(αc) geometry adopted from Fig 4.7 and Fig 4.11: L1= 2.809mm and αc = 4 degree

Effect of Integrated bleed and cowl deflection angle configuration on Mach contours:-

With an emphasis on Mach contours, the study explores how different cowl deflection angles affect scramjet inlet performance. Results show that a specific cowl deflection greatly lessens the separation caused by lip-shock on the inlet ramp and lessens the impact of shock at the isolator section, which improves pressure recovery at the isolator exit. This conclusion is reinforced by accompanying plots.

A comparative analysis of the situations with and without cowl deflection angles reveals significant differences in flow separation patterns, such as shoulder separation, cowl tip separation, and isolator separation. Using the cowl deflection angle at the inlet throat lessens the effect of the shoulder separation, thus changing the shock train behavior that follows in the isolator section. More examination of the effect of various cowl deflection angle configurations on pressure recovery at the isolator exit provides valuable information for performance optimization of the inlet.

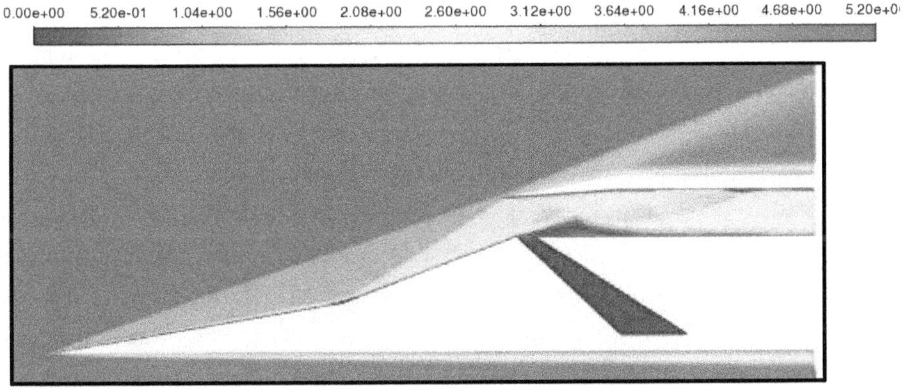

Fig 4.16: Mach contour representing the flow physics for inlet with best bleed gap case and best cowl angle deflection case integrated

Effect of Integrated bleed and cowl deflection angle configuration on surface pressure distribution on ram:-

Typically, a specific number of shocks perform the compression in the scramjet inlet. Figure 4.17 illustrates the Shock Boundary Layer Interactions within the inlet-isolator body by showing the non-dimensionalized pressure distribution on the ramp surface for different cowl deflection angle configurations.

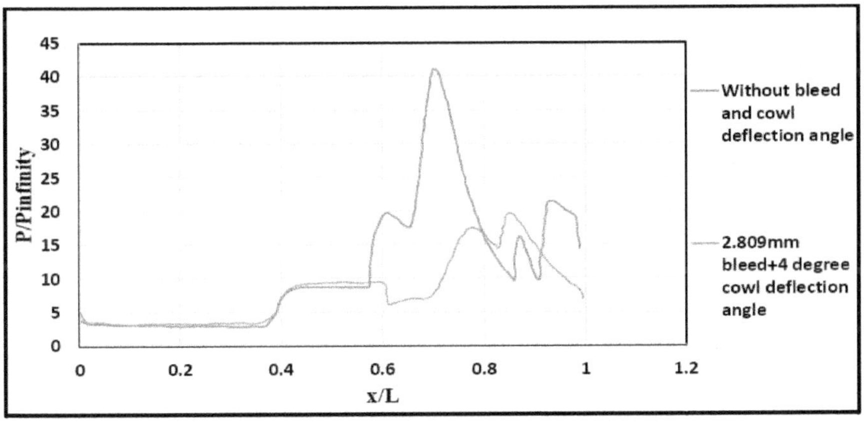

Fig 4.17 Non-dimensionalized pressure distribution on ramp surface combined best case bleed and best case cowl geometry

The plot depicts the variation of surface pressure distribution along the length (x/L) of the scramjet inlet. Specifically, the segment from position 0 to 0.4 corresponds to the surface pressure distribution at the first ramp (10-degree ramp), and the segment from 0.4 to 0.6 represents the surface pressure distribution at the second ramp (22-degree ramp). A discernible increase in pressure occurs beyond x/L 0.6, which is suggestive of the pressure distribution at the isolator section.

We can see that pressure loss results from an abrupt rise in the surface pressure distribution at the isolator section. The diminished efficiency of the scramjet inlet can be attributed to three types of flow separation: shoulder separation, cowl tip separation, and the third separation occurring at the isolator. Our aim is to the reduce this pressure loss and thereby increasing the pressure recovery on the inlet which in turn increases the overall efficiency of the inlet which is obtained and is shown in the results.

Effect of Integrated bleed and cowl deflection angle configuration on the total pressure recovery at isolator exit:-

The comparison of total pressure recovery for cowl deflection angle configuration and integrated bleed in free flow at the exit is shown in Figure 4.18. The overall pressure recovery at the Scramjet Inlet's outlet is shown in this plot. The ratio of the initial stagnation pressure when the air enters the inlet to the stagnation pressure at the inlet's outlet is known as total pressure recovery (π). It functions as a gauge for the overall pressure loss brought on by the compression process.

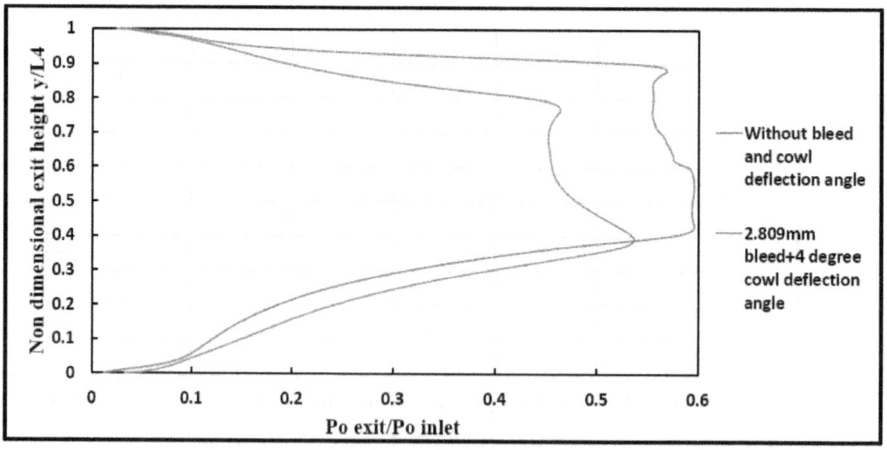

Fig 4.18 Comparison of total pressure recovery combined best case bleed and best case cowl geometry

The y-axis is represented by the length L4 (refer Fig 4. 6) which is non dimensionalized for this study, and the pressure recovery is measured at this location for the cases with bleed and without bleed. This metric is crucial for gauging the total pressure loss incurred during the compression process. The efficiency of total pressure, a key parameter, is notably impacted by interactions between shock waves and the boundary layer. Hence, pressure recovery emerges as a critical factor in assessing the pressure losses within the inlet- isolator system, contributing to the overall enhancement of the inlet's efficiency.
Below are the pressure recovery results for the validation case and the cases with cowl deflection angles.

Table 4.7: Percentage increase in total pressure recovery for different cases

Case	Total pressure recovery Π	Percentage increase in total pressure recovery
Validation case	0.287144	NA
2.809 mm bleed + 4 degree cowl deflection angle	0.382865	33.33554%

4.7 RESULTS AND DISCUSSION

In brief, the evaluation of the scramjet inlet-isolator performance was mostly about total pressure efficiency (πc), which is a pivotal parameter that basically accounts for the total pressure lost during compression. This lost total pressure can be due to many different causes, such as shockwaves, boundary layer interactions, or viscous effects. Total pressure efficiency, expressed as π, was obtained from the comparison of the isolator exit stagnation pressure and the freestream stagnation pressure.

Table 4.8: Cumulative results of performance of inlet for various cases

Sl. No.	Case	Pressure recovery Π_c at outlet (L4)	Percentage increase in pressure recovery
1.	Base case	0.287144	-
2.	1.4045mm	0.296668	3.3168%
3.	2.809 mm	0.3465767	20.697%

4.	4.2135 mm	0.3416698	18.989%
5.	5.618 mm	0.3288346	14.519%
6.	1 degree	0.302742	5.4321%
7.	2 degree	0.348722	21.4449%
8.	3 degree	0.308617	7.4781%
9.	**4 degree**	**0.379791**	**32.2649%**
10.	5 degree	0.34167	18.989%
11.	2.809mm bleed + 4 degree cowl deflection angle	0.382865	33.33554%

The introduction of a 2.809mm bleed substantially raised the pressure recovery thus led to an efficiency enhancement of 20.697%. Likewise, a 4-degree cowl deflection angle usage caused a considerable efficiency increase of 32.2649%. The results therefore show that both the bleed and the cowl deflection angle were very effective in the case of the scramjet inlet.

On the other hand, a combination of the 2.809mm bleed and 4-degree cowl deflection angle in the base geometry resulted in a relatively small efficiency gain of 33.33554% with respect to the base case. It was even more surprising that, when comparing this combined configuration with the 4-degree cowl deflection angle alone, the former was less efficient. The fact that the 4-degree cowl deflection angle can be considered as the top most effective in the enhancement of total pressure efficiency among ten different enhanced cases is therefore an implication of this observation.

CHAPTER 5
NUMERICAL SIMULATION OF SCRAMJET COMBUSTOR

5.1 Introduction

Supersonic Combustion Ramjets, or Scramjets, are a type of engine that attains hypersonic speeds (beyond Mach 5) through the use of air-breathing propulsion. Scramjets, unlike regular jet engines, operate at much higher speeds and do not have rotating parts such as turbines. The working of a scramjet is centered around the compression of air that is drawn into the engine. This is done by using the speed of the plane. Once the air is hot and compressed, fuel is mixed with the flow and set on fire, thus resulting in supersonic combustion.

The combustion process in scramjets is sustained by the high-speed air that is passing through the engine. The combustion generates a thrust that moves the aircraft forwards. Scramjets are designed to be operable at very high speeds and altitudes where the air is thin and a normal jet engine will not be able to work. Moreover, the faster the vehicle, the more the scramjet is likely to become efficient as the high-speed air passing through the engine will enhance the combustion process.

In general, scramjets are extremely complicated and highly sophisticated engines which demand accurate engineering and materials to be able to function under such harsh conditions. They provide a possibility for very fast travelling within the atmosphere and could be utilized in the next generation of aircraft and space vehicles for various purposes, such as military and civilian transport.

The high-speed air is compressed at the scramjet's inlet by the ram effect.
Fuel: most commonly hydrogen or a hydrocarbon is sprayed into the supersonic airstream where it is mixed, ignited, and consumed. The desired thrust is produced by expelling the high enthalpy gases through a nozzle.

In order to produce a highly efficient scramjet combustor, changes are introduced in the combustor by the addition of various auxiliary devices such as struts and cavities. The covered here study focuses on the usage of different flame holders and cavities in combustion and how it affects the process.

5.2 Design Methodology

The ANSYS design methodology begins with problem definition, where the engineering challenge is clearly identified and objectives are established. This is followed by the creation of a 2D or 3D model, either built within the software or imported from external CAD tools. Once the geometry is prepared, mesh generation is carried out by discretizing the domain into finite elements suitable for numerical analysis. Appropriate material properties are then assigned, and boundary conditions such as loads, constraints, and operating conditions are defined to accurately represent the physical problem. Subsequently, the solver setup is configured by selecting suitable numerical models and solution methods based on the governing physics.

After setting up the simulation, the solution phase is executed to obtain numerical results. These results are then examined during post-processing, where key parameters are analyzed and visualized using plots and contours. Based on these insights, optimization and iteration are performed to refine the design and improve performance. The simulation results are further subjected to validation by comparing them with experimental data or established benchmarks to ensure reliability. Finally, the entire workflow is systematically recorded during documentation, ensuring transparency, reproducibility, and clarity of the simulation process.

5.3 Geometry and Domain

Geometry:

DLR model schematic is used as shown below in Figure 5.1. The combustor is of height of 50mm at the entrance and 62mm at the exit. The divergence is directed from the top wall after 100mm with a 3 degree. The strut is placed at point 77mm away from the inlet and 25mm from the bottom wall and the bottom wall is of length 340mm. The current investigation uses three distinct boundary types: inflow, outflow, and fixed walls. The outlet boundary is a pressure outlet boundary type, and all input boundaries and fuel—are regarded as pressure inlets. The permanent walls are subjected to the no-slip requirement. Figure 2 displays a typical grid, the computational domain, and the boundary types used in the present investigation. Experimental results for the DLR model have been extracted at different locations (X=125 mm, 157 mm, 233 mm, 275 mm) as indicated in Figure 1 to cover the mixing, combustion, and after-combustion zones. Table 1 tabulates the airstream and hydrogen input conditions.

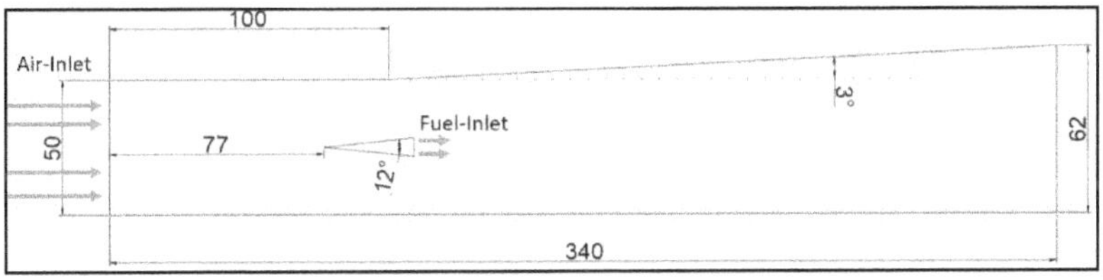

Fig.5.1 Geometry of standard DLR model

Meshing and Boundary conditions:

The mesh type used for this model is global sizing of 3e-04m and additionally, the inflation is applied at walls to get the wall Y+ less than 2 and get accurate results on walls as shown in Figure 5.2. The first layer thickness applied is 1e-03mm.

The mesh count is: -

Number of elements: - 80,000. Number of nodes: - 2,20,000.

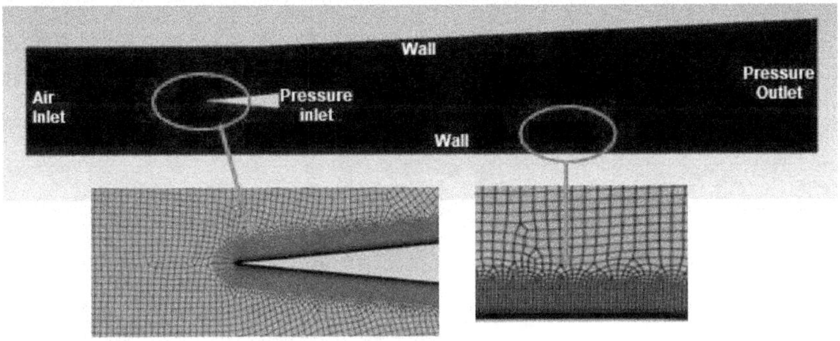

Fig 5.2: Meshing and Boundary conditions

In the present work, the RANS equations are numerically solved using the commercial program ANSYS-FLUENT. When modelling chemically reactive flow systems, the turbulence model is essential. The turbulence models used determine the correctness of the flow problem. Using the finite rate/eddy dissipation combustion model and the SST k-omega turbulence model, the numerical analysis was carried out in a 2-D planar steady state. Fuel and oxygen react instantly to form products as soon as they come into contact, according to the Eddy Dissipation Model (EDM), which is based on the premise of infinitely fast chemistry. This assumption relies on the idea that the chemical reaction rate far exceeds the speed of the physical flow processes within the system. [135-135].

According to this model, the rate at which the reactants mix at the molecular level determines the reaction rate rather than the speed of the chemical reactions. The duration for this mixing operation is essentially determined by the turbulent flow structures that are present in the system. Turbulence is very important to the mixing of the reactants as it breaks them down into smaller scales, thus the interface where the chemical reactions can take place is increased. In the EDM framework, the reaction rate is associated with a "mixing time" which is obtained from the turbulent kinetic energy (k) and its dissipation rate (ϵ). These parameters influence the rate and efficiency of the fuel and oxidizer's mixing and reaction by indicating the magnitude and the intensity of the turbulence in the flow. This model works very well in conditions where mixing-limited combustion is the case, for example, in high-speed flows, and it makes the analysis simple by not requiring the solution of detailed chemical kinetics.

The finite-rate eddy dissipation turbulence model explains the changes of turbulence and chemical processes that are coupled dynamically. The study employs a single-step chemical reaction, where hydrogen reacts with oxygen to produce water. A single-step chemical reaction can evaluate all parameters for scramjet engine combustion [87, 106]. Using this approach, the manuscript accurately predicts performance metrics while greatly decreasing computational costs.

$$2H_2 + O_2 \rightarrow 2H_2O \tag{1}$$

The ideal gas was used for both the air and fuel streams. The following standard equations mass, momentum, energy conservation and turbulence modelling are solved, along with finite rate species transport equations, whereas turbulence chemistry interaction is modelled by eddy dissipation model and solved in ANYSYS FLUENT solver.

Continuity equation:

$$\frac{\partial \rho}{\partial t} + \frac{\partial}{\partial x_i}(\rho u_i) = 0 \tag{2}$$

Momentum equation:

$$\frac{\partial}{\partial t}(\rho u_i) + \frac{\partial}{\partial x_i}(\rho u_i u_j) = -\frac{\partial P}{\partial x_i} + \frac{\partial \tau_{ij}}{\partial x_j} \tag{3}$$

Energy equation:

$$\frac{\partial}{\partial t}(\rho E) + \frac{\partial}{\partial x_j}\left(u_j(\rho E + P)\right) = \frac{\partial}{\partial x_j}\left(\tau_{ij}u_j + k_{eff}\frac{\partial T}{\partial x_j}\right) + S_h \tag{4}$$

The turbulence kinetic energy (k) and specific dissipation rate (ω) are obtained by solving the transport equations defined for k and ω. These transport equations are defined by the production caused by mean velocity gradients, dissipation caused by turbulence, and diffusive transport as shown in equations 5 and 6 [50].

$$\frac{\partial}{\partial t}(\rho k) + \nabla_i(\rho u_i k) = \frac{\partial}{\partial x_j}\left(\Gamma_k \frac{\partial k}{\partial x_j}\right) + G_k - Y_k + S_k \tag{5}$$

$$\frac{\partial}{\partial t}(\rho \omega) + \nabla_i(\rho u_i \omega) = \frac{\partial}{\partial x_j}\left(\Gamma_\omega \frac{\partial \omega}{\partial x_j}\right) + G_\omega - Y_\omega + S_\omega \tag{6}$$

Here, G_k represents the turbulence kinetic energy generation due to mean velocity gradients. G_ω represents the generation of ω. Γ_k and Γ_ω represent the effective diffusivity of k and ω, respectively. Y_k and Y_ω represent the dissipation of k and ω due to turbulence. All of the above terms are calculated as described below. S_k and S_ω are user-defined source terms.

Table 5.1: Boundary conditions of the air inlet and fuel inlet

Fluid	Mach Number	Pressure (10^5Pa)	Temperature (K)	Density (Kg/m^3)	Y_{O_2}	Y_{N_2}	Y_{H_2O}	Y_{H_2}
Air (g)	2	1	340	1.002	0.232	0.736	0.032	0
H_2 (g) (Fuel)	1	1	250	0.097	0	0	0	1

5.4 Validation Plot

In the case of reacting flow within the supersonic combustor, a qualitative comparison of the flow field is carried out between the computed density contours obtained from the experimental shadowgraph images reported by Waidmann et al. [83]. The density contour plots provide insight into the shock structures, mixing zones, and combustion-induced variations in density, while the shadowgraph technique captures flow features such as shock waves, expansion fans, and regions of density gradients in the experimental setup. By carrying out this comparison, it is possible to evaluate the capability of the numerical model to capture the essential physical phenomena that have been observed in the experiments, thus validating the accuracy of the numerical approach used in the simulations. Boundary conditions of the air inlet and fuel inlet are shown in Table 5.1.

The static pressure distribution along the bottom wall of the combustor is compared with the experimental data provided by Waidmann et al. [83] and the numerical simulations performed by Ritesh et al. [84]. Both reacting and non-reacting flow field have been discussed. Non-reacting flow field can be studied with fuel injection without combustion or without fuel injection. As can be seen from Figure 3 (a) and (b), the computational results are in good agreement with the experimental data and the previous numerical simulations, thus validating the accuracy of the numerical approach used in the simulations. In addition, the density contour plot obtained from the present numerical simulations is compared with the experimental Schlieren photograph, as shown in Figure 3 (c) and (d). It can be seen that there is a fair level of qualitative agreement between the two, thus validating the accuracy of the simulation in capturing the essential physical phenomena observed in the experiments. The results from the numerical simulations are in good agreement with the experimental data, thus validating the accuracy of the solver.

(a)

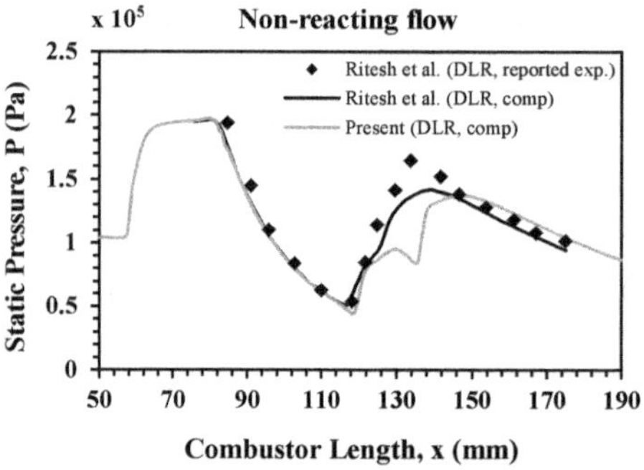

(b)

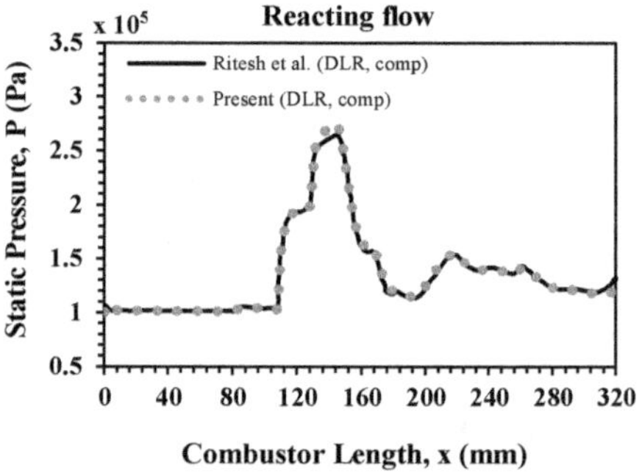

(c)

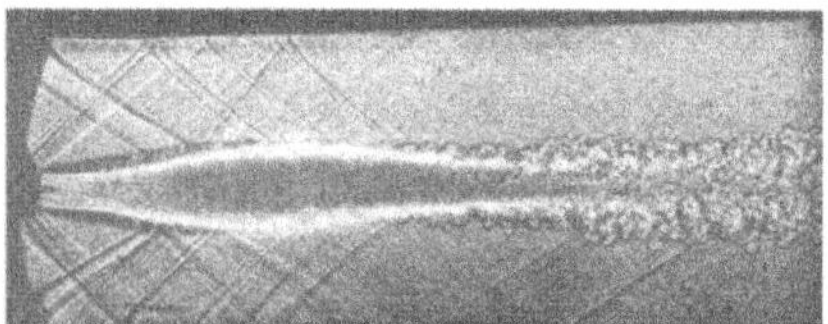

(d)

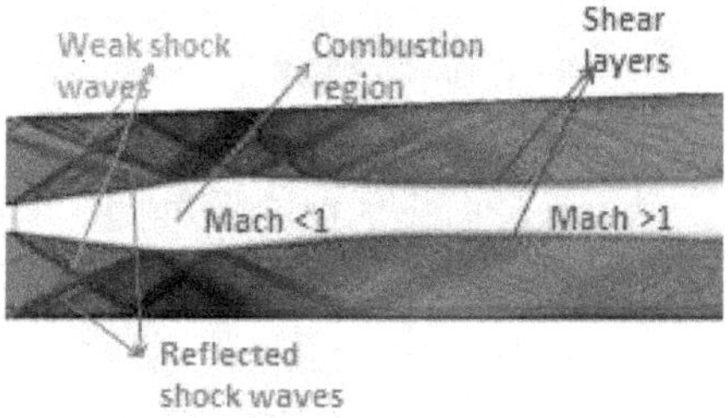

Figure 5.3. Static pressure distribution along the bottom wall of the combustor for (a) non-reacting flow and (b) Reacting flow in comparison with results by Ritesh et al. [83]. (c) The experimental schlieren image of density from Waidmann et al. [84]. (d) Computed density contour of reacting flow (present study)

5.5 Grid Independence Test

Based on the qualitative analysis, a quantitative validation and grid independence test were performed using the static pressure and temperature distributions on the combustor wall and chosen flow planes. The global cell size was changed from 0.818 mm to 0.44 mm, and the wall inflation layer thickness was adjusted to keep the wall Y^+ values below unity.

As can be seen from the corresponding results plotted in Figs. 4 and 5, considerable variation is obtained in the solution on coarse meshes; however, convergence is obtained after nearly 100,000 elements, with minor changes as the mesh is further refined. The computed results were validated against the experimental data reported by Waidmann et al. [83] as well as the numerical predictions reported by Ritesh et al. [84]. Accordingly, a fine structured mesh of 0.5 mm element size was chosen for further simulations.

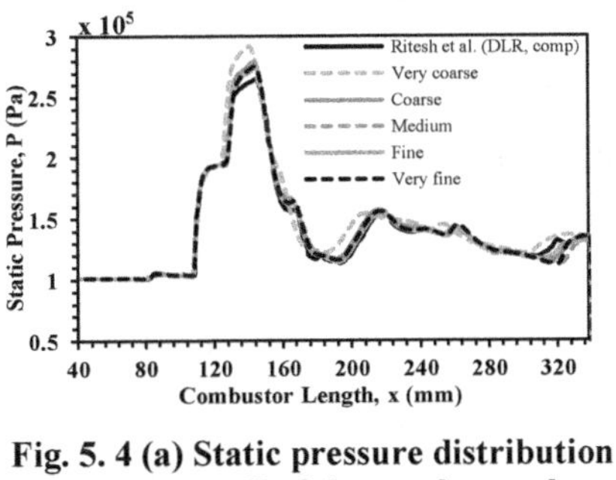

Fig. 5. 4 (a) Static pressure distribution along bottom wall of the combustor based on different mesh size.

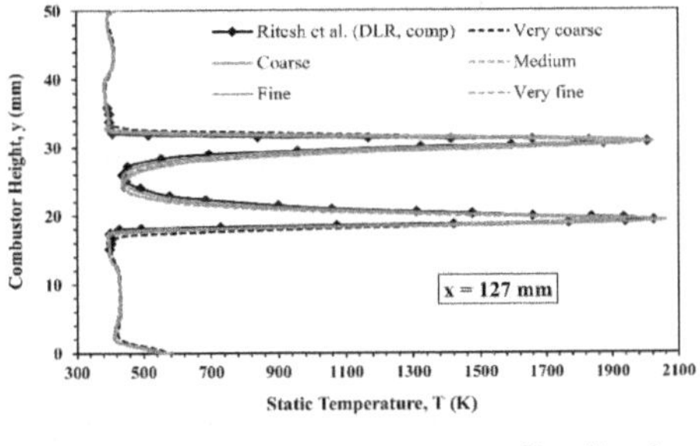

Fig. 5. 4 (b) Static temperature distribution along the height of the combustor at a distance of 127mm from the combustor inlet based on different mesh size.

5.5.1 Numerical uncertainty assessment

Numerical solvers are generally iterative in nature, therefore the solution does not converge to an exact value but to a residual tolerance, leaving a small numerical error. This uncertainty needs to be minimized before analyzing any numerical results. In the present study, steady-state simulations are performed, using density-based solver with a pseudo-time iteration. Therefore, numerical errors are primarily associated with spatial discretization and iterative convergence, rather than long-time accumulation of physical time integration errors. Convergence of the solution is ensured by monitoring residual reduction and stabilization of key flow variables such as wall pressure, temperature distribution and species mass-fractions [54, 55]. In addition, sensitivity to pseudo-time step (Courant number) was examined, and no significant variation in the solution was observed within the selected range, confirming that iterative numerical errors are well controlled. Grid convergence analysis is further carried out using the Grid Convergence Index (GCI) method to quantify discretization errors. The procedure followed is based on the methodology reported by Ritesh et al. [84]. The present geometries are evaluated for their numerical uncertainty using GCI method with basic steps described as below.

Φ is considered as a flow property such pressure, temperature, density, etc. Area-weighted-average of static temperature at the outlet of the combustion is chosen as Φ. Three grid, coarse, medium and fine was created for the calculation. Using this flow property, difference between two successive grid is calculated as:

$$\epsilon_{12} = \Phi_1 - \Phi_2 \quad (5)$$

$$\epsilon_{23} = \Phi_2 - \Phi_3 \quad (6)$$

Next, approximate order of convergence is calculated using:

$$P = \frac{1}{ln(r)} \left| ln\left(\frac{\epsilon_{12}}{\epsilon_{23}}\right) \right| \quad (7)$$

Where r is the average refinement ratio. It is obtained from the average of r_{12} and r_{23} which is calculated as follows:

$$r_{12} = \sqrt{\frac{N_2}{N_1}} \quad (8)$$

$$r_{23} = \sqrt{\frac{N_3}{N_2}} \quad (9)$$

Where N_1, N_2 and N_3 are the cell count of coarse, medium and fine grid respectively. Average refinement ratio and order of convergence are then combinedly use to extrapolate the flow property using Richardson extrapolation method given by:

$$\Phi_{ext}^{12} = \frac{r^P \Phi_2 - \Phi_1}{r^P} \quad (10)$$

$$\Phi_{ext}^{23} = \frac{r^P \Phi_3 - \Phi_2}{r^P} \quad (11)$$

Followed by this, we move to end of the calculation by calculating the relative error, e_a and finally the GCI.

$$e_a^{12} = \left|\frac{\Phi_2 - \Phi_1}{\Phi_2}\right| \tag{12}$$

$$e_a^{23} = \left|\frac{\Phi_3 - \Phi_2}{\Phi_3}\right| \tag{13}$$

$$GCI_{12}(\%) = F_s\left(\frac{e_a^{12}}{r^P - 1}\right) \tag{14}$$

$$GCI_{23}(\%) = F_s\left(\frac{e_a^{23}}{r^P - 1}\right) \tag{15}$$

Where F_s is the safety factor as recommended for studies which involve three grids, to ensure conservative error estimation.

Table 5.2 GCI study of cavity and non-cavity-based configurations

Parameter	Single strut			Dual strut		
	No cavity	Ramp cavity	Step cavity	No cavity	Ramp cavity	Step cavity
Φ_1 (coarse)	871.0042	883.7812	880.0273	1199.5524	1219.9831	1218.8054
Φ_2 (medium)	791.9485	808.4439	803.8511	1113.4911	1134.7751	1133.7332
Φ_3 (fine)	766.4258	782.1855	776.5624	1087.136	1107.0591	1106.5777
ϵ_{12}	79.0557	75.3373	76.1762	86.0613	85.208	85.0722
ϵ_{23}	25.5227	26.2584	27.2887	26.3551	27.716	27.1555
r_{12}	1.1547	1.1724	1.1747	1.1428	1.1782	1.1747
r_{23}	1.118	1.1253	1.1189	1.1093	1.1093	1.1189
R	1.13635	1.14885	1.1468	1.12605	1.14375	1.1468
P	8.8450	7.5957	7.4946	9.9683	8.3617	8.3367
Φ_{ext}^{12}	510.7496	500.4068	488.5978	746.1445	737.9456	744.6840
Φ_{ext}^{23}	510.7496	500.4068	488.5978	746.1445	737.9456	744.6840
e_a^{12}	0.0998	0.0932	0.0948	0.0773	0.0751	0.0750
e_a^{23}	0.0333	0.0336	0.0351	0.0242	0.0250	0.0245
GCI_{12} (%)	5.9491	6.2322	6.6121	4.2646	4.5248	4.3978
GCI_{23} (%)	1.9846	2.2451	2.4519	1.3376	1.5087	1.4383

With the GCI analysis of the present 6 cases as described in table 3 shows error between coarse and medium mesh as 5.9491-6.2322 % in case of single strut configuration and 4.2646-4.5248 % in dual strut configuration. On the other hand, GCI for medium and fine mesh is found to be much lower with range of 1.9846-2.4519 % in single strut configuration and 1.3376-1.5087 % in dual strut configuration. With this understanding, fine mesh of 0.5 mm element size was chosen for all the further discussion.

5.6 Novel Geometries

The current study examines the effects of different cavity angles and dual-strut configurations to improve the performance of scramjet combustors. Figure 5.5 displays the geometric configurations from the analysis of six distinct cases. Case 1 features a typical DLR model of the flame holder, which is taken as the baseline and is used without any modifications. After that, the following cases are changed by the addition of the cavities and the dual-struts to the original model. It is noted that the cavity generates a stable recirculation zone, which acts as a continuous source of hot products and radicals, enabling reliable ignition and flame anchoring in the high-speed supersonic flow.

This recirculation region increases the effective residence time of the fuel–air mixture and promotes sustained combustion despite the short flow timescales. Consequently, the cavity-based flame holder significantly improves flame stabilization and overall combustion efficiency.

In case 2, the bottom wall of the combustor has a 45-degree cavity at a distance of 120mm from the air inlet, as shown in Figure 5. The 90-degree cavity is added in Case 3. In Case 4, the single strut of the baseline model is replaced with the dual-strut, and the exact dimensions are also given in Figure 5. At last, the Cases 5 and 6 combine the dual-struts with the above-mentioned cavities, i.e., the 45-degree and 90-degree ones, respectively.

Such a sequence of experimental rigs makes it possible to comprehensively evaluate the impact of strut geometries and cavity angles on the performance of a scramjet combustor.

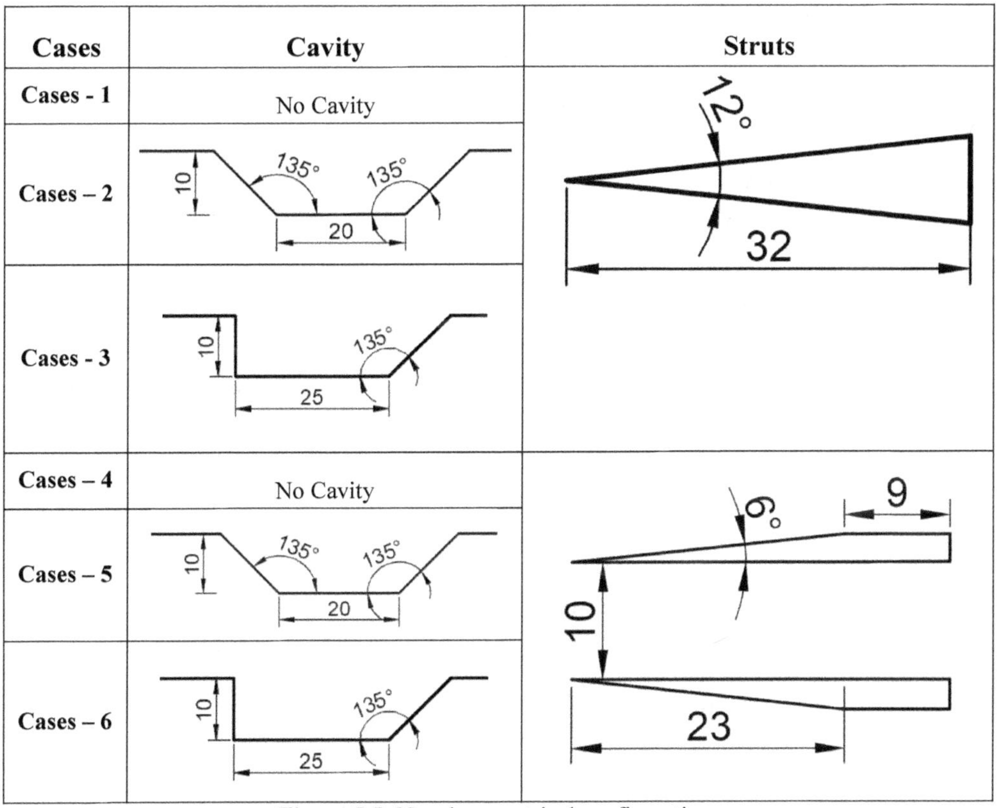

Figure 5.5: Novel geometrical configurations

5.7 Results and Discussions

The goal of the research is to understand the influence of a dual-strut design on the operating parameters of the scramjet combustor. The purpose of the study is to understand how the dual-strut configuration influences the major performance parameters such as fuel-air mixing, combustion efficiency, and flow dynamics of the combustor.

First, computational simulations of cavity-based scramjet combustors with cavity angles of 45 and 90 degrees were performed to identify the effect. Subsequently, numerical simulations were conducted on the dual-strut scramjet combustor.

After that, the dual-strut and cavity were combined to investigate their combined effect on the performance of the scramjet combustor.

5.7.1 Velocity Streamlines

A scramjet combustor's strut aids in creating shock waves inside the combustor, which lowers fluid velocity and increases air-fuel mixing time. Similarly, the cavity and dual-strut help in creating vortices and turbulence inside the combustor and help in better mixing of air-fuel mixture. To visualize the vortices and re-circulation zones formed inside the combustor, the velocity streamline contours are observed. In case-2 and case -3 due to the cavity, the vortices are generated which helps in air-fuel mixing as shown below in Figure 5.6. Even in dual-strut (case-4) a few vortices are observed at the upper and lower wall along with a wall of cavity. These vortices can also be observed in case-5 and case-6 with case-5 possessing a higher number of re-circulation streamlines as shown below in Figure 5.6.

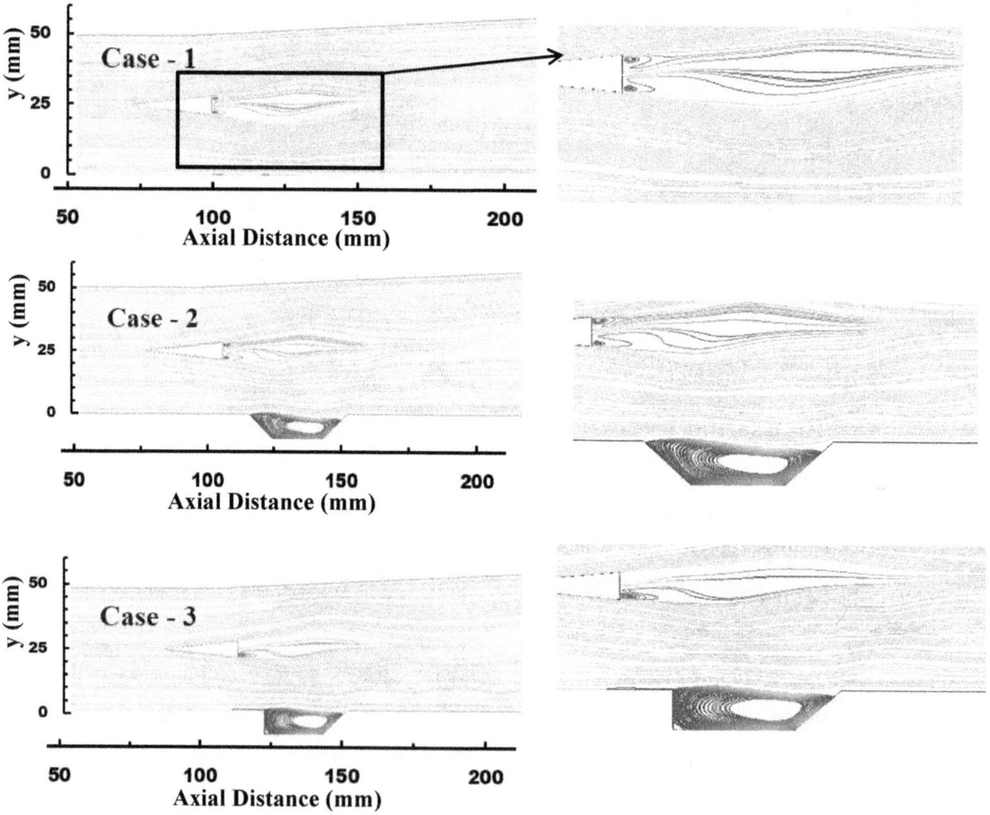

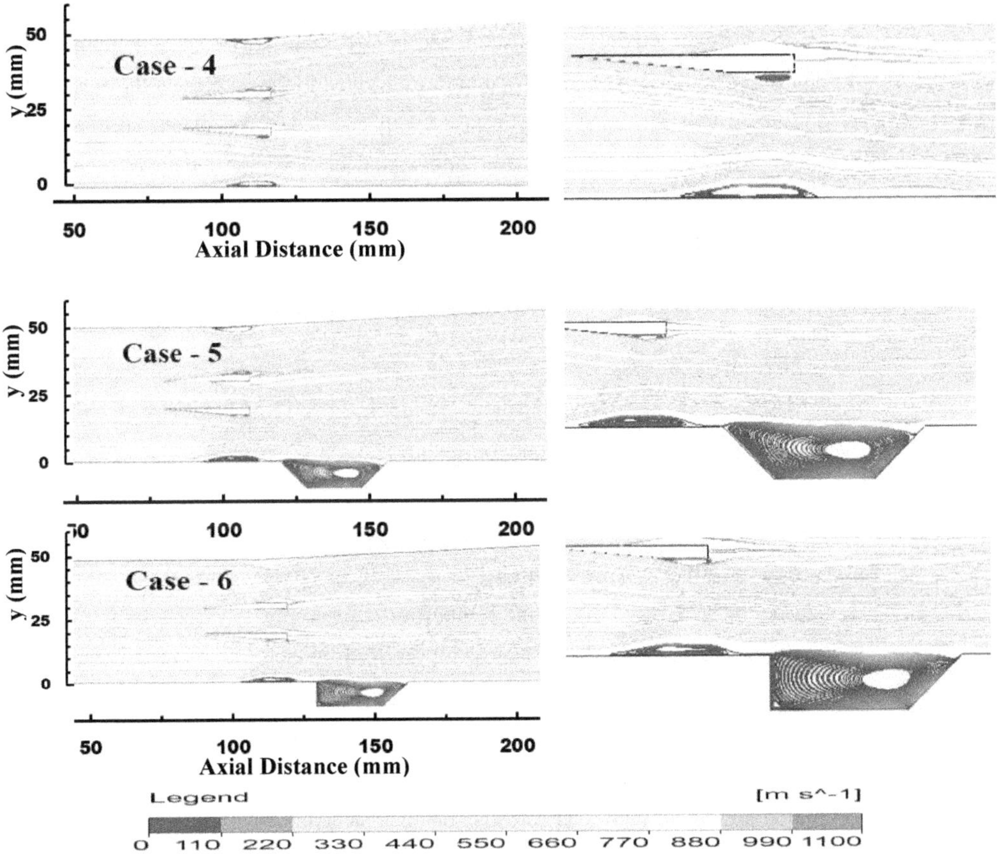

Figure 5.6: Velocity Streamline Contour of all 6 cases.

5.7.2 Velocity Profile

In scramjet combustors, airflow velocity plays a crucial role in optimizing the fuel-air mixing ratio. However, the significant velocity difference between the incoming air and fuel makes achieving effective mixing a challenging task. To enhance mixing performance, it is essential to reduce the airflow velocity, thereby increasing the residence time for air-fuel interaction. This study investigates the implementation of cavities and dual-struts to generate vortices that reduce airflow velocity and improve mixing efficiency. Velocity measurements are conducted for all six configurations at various locations along the combustor length (X = 125 mm, 157 mm, 233 mm, and 275 mm the lengths are also indicated in Figure 5.1) and are shown in Figure 5.7.

Experimental results for the DLR model were gathered at the four locations mentioned above to cover the mixing, combustion, and after-combustion zones. The points (X = 125 mm, 157 mm, 233 mm, and 275 mm were selected as they represent the areas of the combustor where the most significant flow phenomena take place, such as [e.g., transition from the injector to the mixing region or interaction with cavity-induced vortices]. These positions provide measuring means for the velocity changes at different stages of flow development and the evaluation of the mixing and combustion process performance.

The results show that cases 2 and 3 had velocities slightly lower than that of case 1. At X = 125 mm, the velocity dropped by 4% for case 2 and by 2% for case 3, which also points to this region as a critical one for mixing.

Additionally, the analysis of the data disclosed that the velocities in cases 4, 5, and 6 were significantly lower as compared to the corresponding values in the combustor. Their velocity at X = 125 mm in the combustion zone dropped by 21%, 16%, and 18%, respectively. It was noticed that among all the measurement points in the combustor, case 4 always exhibited the lowest velocity. On the other hand, at all the locations, the velocities in case 5 were higher than those in two-strut configurations, while at X = 157 mm and 233 mm, case 2 had the highest velocities when compared to other configurations. In all dual-strut cases, velocities reached 25% higher in the combustion zone than single-strut cases, mainly due to better initial mixing at X=125mm and faster combustion at X=157mm. There were also similar velocities even further down the combustor length, which indicated more complete combustion before X=233mm.

From the results, it is evident that the inclusion of cavities and dual-struts inside scramjet combustors leads to the reduction of airflow velocities, which is a necessary condition for the improvement of air-fuel mixing. The lowest velocities were achieved by case 4 and case 5 yielded good results at higher velocities, thus contributing important compromises in the decisions on design strategies that inherently maximize combustion efficiency. In the future, the research will be focused on the implications of the data for the efficiency and performance of scramjet.

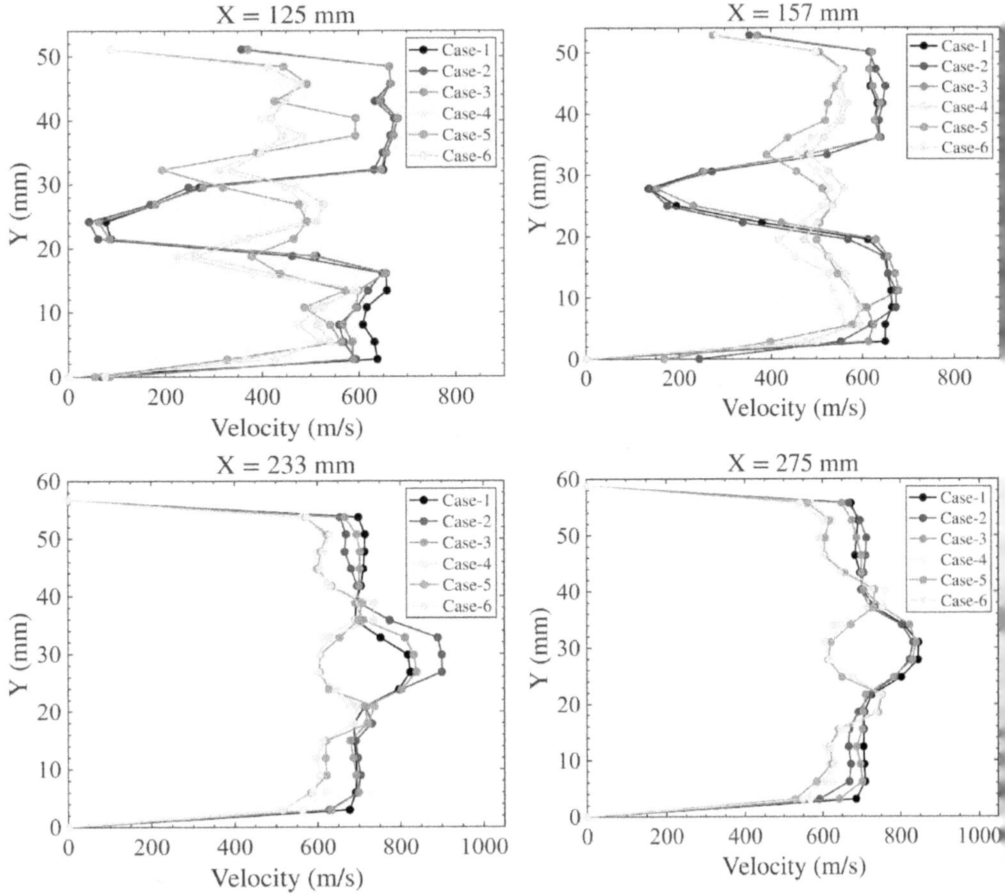

Figure 5.7: Velocity profiles at X=125mm, X=157mm, X=233mm and X=275mm.

5.7.3 Mach Contour

In the last part, we emphasized how important airflow velocity is to the performance of a scramjet combustor. We are now using qualitative analyses of Mach contour rendering to demonstrate these dynamics more vividly and to reflect the change in speed of the airflow for a different set of configurations.

Firstly, comparison of single strut and dual strut configurations through the Mach contour rendering reveals that there are considerable variations in airflow velocity. Secondly, one can point out practically no difference in airflow velocity as the cavities are introduced to the scramjet combustor model, and undoubtedly below in Figure 5.8, the velocity difference between the dual and single strut configurations is significant. In detail, the dual strut reveals lower Mach numbers, which corresponds to the lighter red colour in the wake region, that is, the area of the mixing process. The lower Mach number, therefore, leads to the reduction of airflow velocity and hence better mixing of the air-fuel mixture.

The difference is so great that only one interpretation is possible, namely, the scramjet combustor in a dual-strut arrangement performs much better than in a single strut configuration. Moreover, this research work offers an abundance of useful information on the effects of geometric adjustments with regard to flow properties, which is able to lead to the dual strut promoting the improvements in combustor performance. The next research phase will investigate the relations of these velocity differences with overall combustion efficiency.

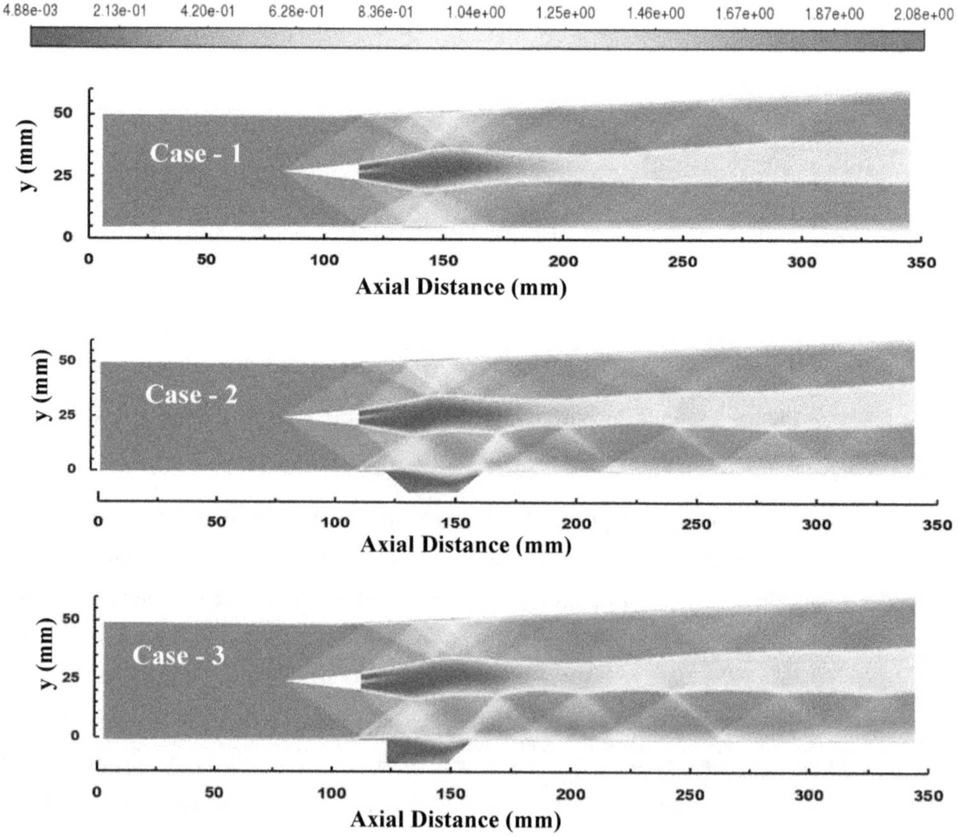

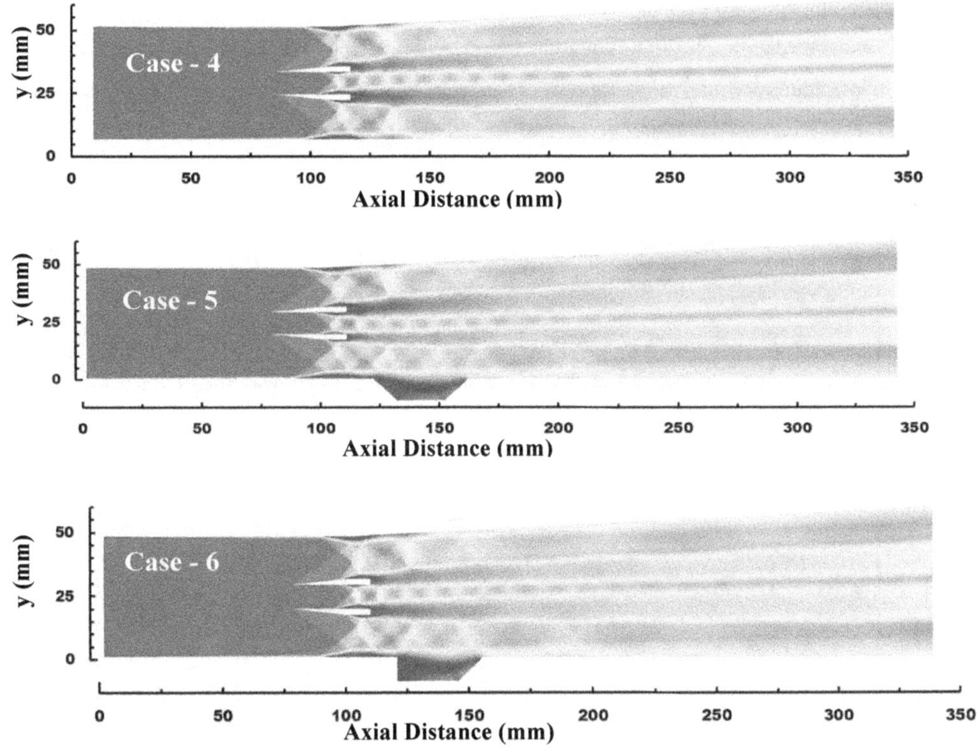

Figure 5.8: Mach Contour of all the 6 cases

5.7.4 H_2 Mass Fraction

The current study examines a single-step reaction mechanism for the production of water (H_2O) through the combustion of hydrogen and oxygen. The contours of the hydrogen mass fraction are plotted in order to clarify the air-fuel mixing processes and reaction dynamics. A sizable amount of hydrogen is injected into the combustor from the strut's center, with an injection area of 1 mm, as shown in Figure 5.9. According to the analysis, the hydrogen mass fraction is higher in the single-strut configuration than in the dual-strut configuration. This observation corresponds with earlier findings, which reported that the lower velocities in the dual-strut cases are beneficial to the mixing performance.

It can be seen from the contours of the hydrogen mass fraction that, because of better fuel-air mixture in the dual-strut cases, less hydrogen is stored within the combustor. Instead, more hydrogen effectively mixes with the air, facilitating earlier and more efficient combustion. This enhanced mixing not only contributes to improved combustion efficiency but also has implications for reducing the overall length of the combustor, as the air-fuel interaction occurs more rapidly. These results emphasize the significance of structural design in optimizing the mixing of fuel and air, thereby enhancing combustion performance in scramjet applications.

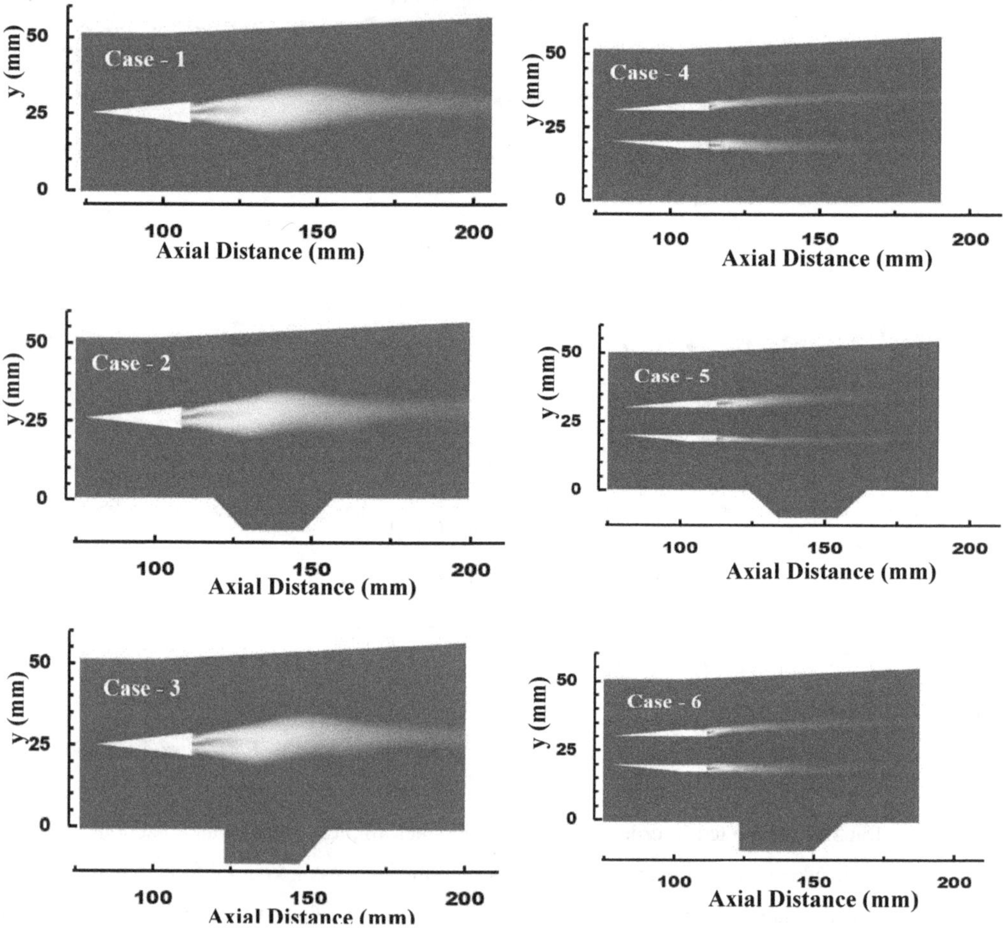

Figure 5. 9: Hydrogen Mass Fraction Contour of all 6 cases.

5.7.5 H_2O Comparison Graphs

Here, the work is to analyze the reaction mechanism of a single-step chemical reaction of the combusting process involving hydrogen and air to produce water (H_2O). We, as the first step to understand the efficiency of air-fuel mixing, had a look at the hydrogen mass fraction. The next thing to do in order to really know how complete the combustion in the combustor is, would be to figure out the products from the reactants, most notably the temperature change due to the reaction. So, in this regard, the mass fraction of H_2O, the main product of the combustion reaction, is used to evaluate the combustion process more thoroughly.

The average H_2O mass fraction is computed for different points in the combustor (X = 125 mm, 157 mm, 233 mm, and 275 mm) for all six configurations. It is apparent from the information that the average H_2O mass fraction for dual-strut models has been significantly elevated compared to that for single-strut models. Namely, the dual-strut configurations at X = 125 mm demonstrate an average increase of 28% in the H_2O mass fraction. This behavior is reflected at other points of the measurements as well, where increments of 7%, 20%, and 12% can be found at X = 157 mm, 233 mm, and 275 mm, correspondingly as shown in Figure 5.10.

It is, therefore, concluded that the H_2O mass fraction in the dual-strut configurations is significantly higher, thus illustrating the increased combustion efficiency. This, in turn, implies that the dual-strut design not only enables the fuel and air to be better mixed but also generates more complete combustion which is the main result, among others, of the overall performance of the combustor being optimized.

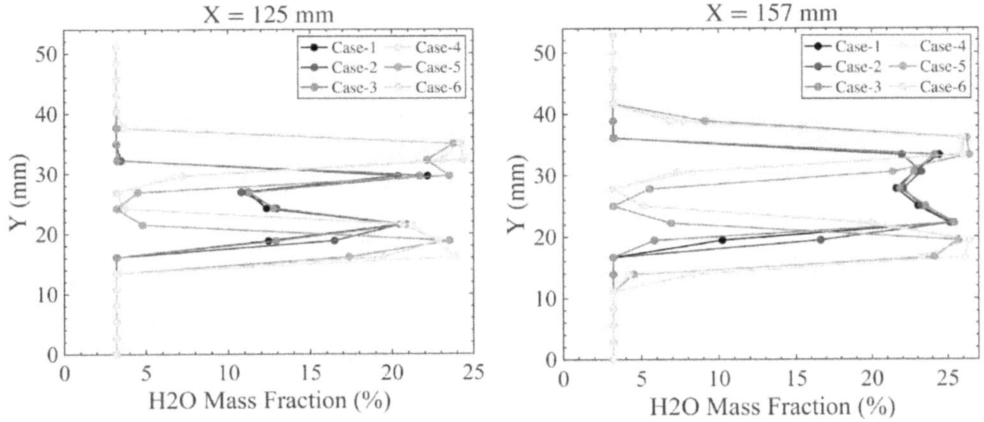

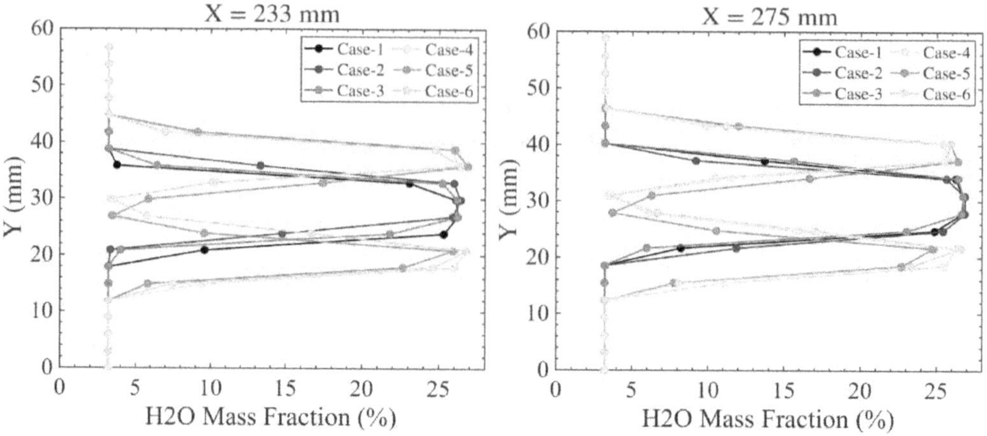

Figure 5.10: H_2O Mass fraction at X=125mm, X=157mm, X=233mm and X=275mm.

5.7.6 Temperature

Temperature plays a crucial role in the burning process and has a major impact on the overall burning efficiency. As a matter of fact, the changes in velocity and H_2O mass fraction at different points in the scramjet combustor were compared and significant trends were identified. Most notably, the dual-strut combustor models exhibit a pretty hefty jump in temperature, while the effect of cavities seems to be negligible. Figure 5.11 displays the temperature increment percentages and the corresponding bar chart of the temperature rises at the 125mm location for the different cases. The average temperature for all six configurations at the 125 mm point is calculated. The first graph shows the percentage changes in temperature from Cases 2 to 6 compared to the Standard DLR model (Case 1). The most dramatic temperature elevation (by about 50%) can be seen for the dual-strut combustor with cavity cases in the fifth and sixth scenarios.

Moreover, the temperature increments of 2.3%, 1%, 34.3%, 36.7%, and 36.4% for Cases 2, 3, 4, 5, and 6, when compared to the Standard DLR model (Case 1) have also been identified by looking at different locations along the combustor. Even Figure 5.12 illustrates the temperature variations calculated and their corresponding high temperatures are observed where dual-strut cases have higher temperature values than single strut cases.

All these temperature increases are indicative of the higher combustion efficiency that the dual-strut models have achieved, especially in Cases 5 and 6, which correspond to the highest temperature values that were recorded. The reason behind it could be the dual-strut device that made the combustion process super-efficient and, hence, optimized the scramjet engine performance.

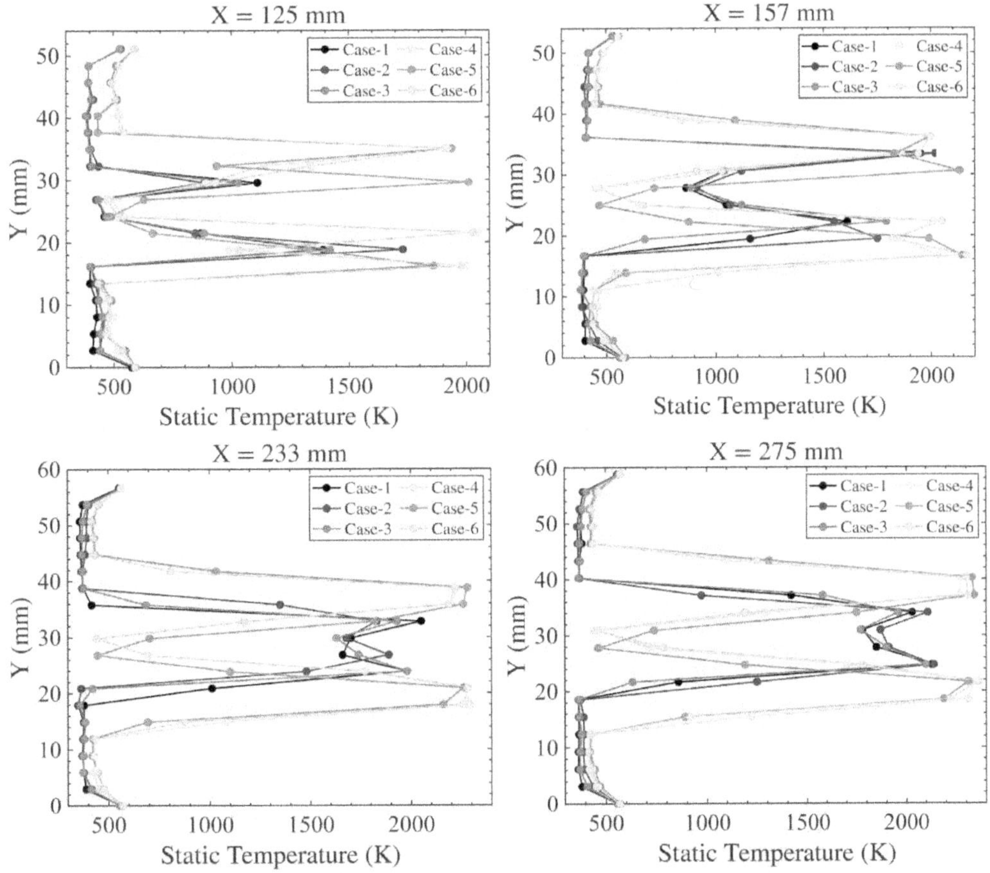

Figure 5.11: Temperature profiles at X=125mm, X=157mm, X=233mm and X=275mm.

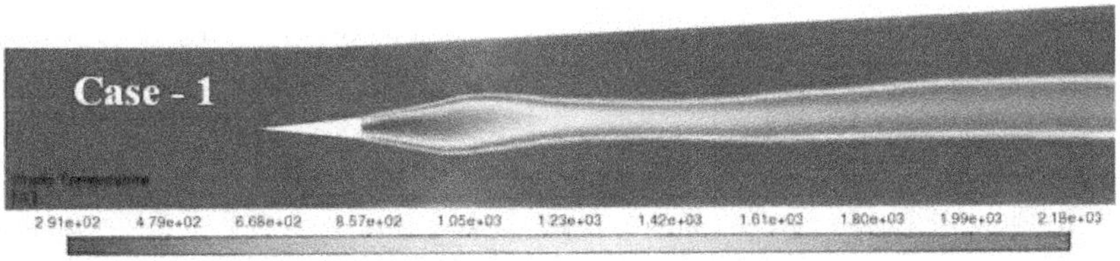

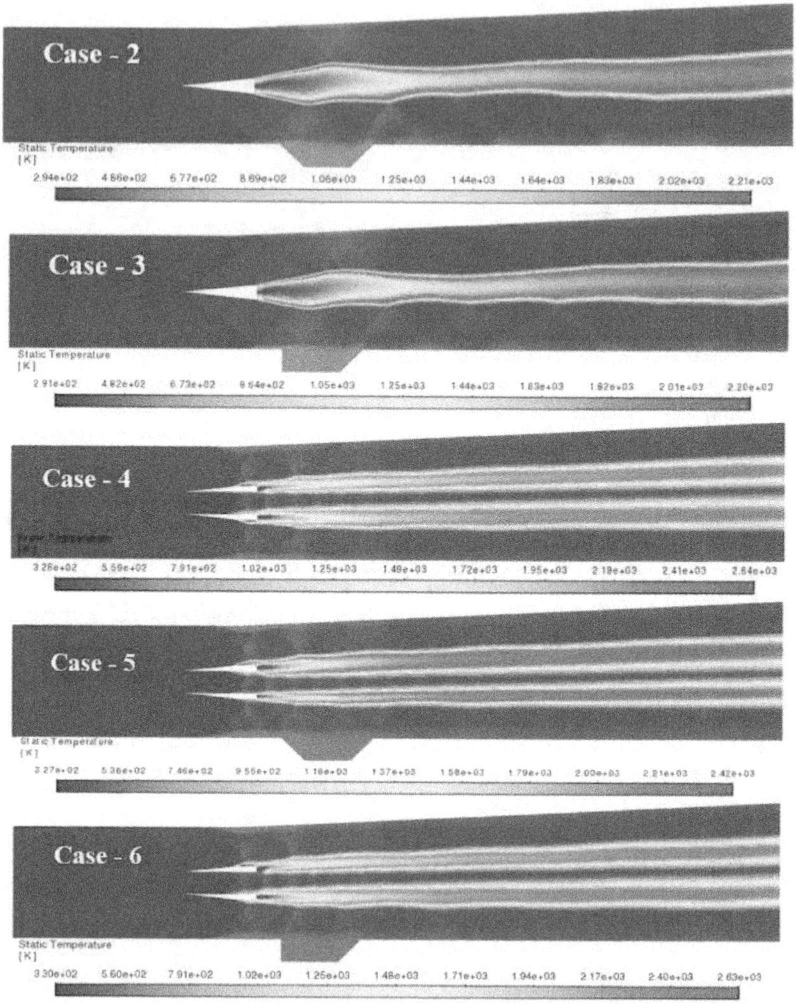

Figure 5.12. Temperature contours of all 6 cases

5.8 Performance parameters

Mixing efficiency and combustion efficiency

The performance parameters, specifically mixing efficiency and combustion efficiency, are critical in assessing the overall effectiveness of scramjet engines. In present work, the performance of six configurations by calculating and comparing their mixing and combustion efficiencies based on numerical computational studies of mass fractions of hydrogen and water (H_2O) are analyzed.

Mixing efficiency was determined using the formula [107]:

$$\text{Mixing Efficiency} = \frac{\text{Missing Mass Fraction of Hydrogen at location}}{\text{Mass Fraction of Hydrogen at location}}$$

$$\eta_{mix} = \frac{\int_A \alpha \rho_{gas} Y_{H_2} u dA}{\dot{m}_{H_2}}, \text{ with } \alpha = \begin{Bmatrix} 1, & \emptyset < 1 \\ 1/\emptyset, & \emptyset \geq 1 \end{Bmatrix} \tag{16}$$

$$\eta_{mix} = \frac{m_{H_2(inj)} - m_{H_2(x)}}{m_{H_2(x)}} = \frac{m_{H_2(inj)}}{m_{H_2(x)}} - 1 \tag{17}$$

To evaluate combustion efficiency, a similar formula is employed which is expressed as [107]:

$$\text{Combustion Efficiency} = \frac{\text{Missing Mass Fraction of Hydrogen at location}}{\text{Injected Mass Fraction of Hydrogen}}$$

$$\eta_{comb}(x) = 1 - \frac{\int (A(x)) \rho_{gas} u Y_{H_2} dA}{\dot{m}_{H_2(inj)}} = 1 - \frac{m_{H_2(x)}}{m_{H_2(inj)}} \tag{18}$$

The mixing and combustion performance rates of all cases are determined for several sections of the combustor as depicted in Figure 5.13. It is important to mention that the dual-strut model reached 100% mixing efficiency far earlier than the single strut configurations. In particular, the dual-strut model achieved 100% mixing efficiency at a distance of 140mm from the initial section of the combustor, or in other words, at a distance of 31mm from the site of fuel injection. In contrast, a single strut model was able to attain that performance at a distance of 250mm from the combustor's initial section or in this case 141mm from the fuel injector. The fact that 100% mixing efficiency can be achieved at such short distances suggests that there will be savings in combustor length.

On combustion efficiency, further comparisons using all configurations also indicate that the single strut models attained a peak rate of 65% whereas the dual-strut configurations were as high as 97%. This is an increase of 32% in combustion efficiency in favour of the dual-strut configurations. These results demonstrate the improved performance of the dual-strut design with respect to the mixing and combustion efficiencies which can be beneficial in the sustenance of scramjet engines. The future studies will focus on these efficiencies and the impact they have on the design of the engine and its functionality.

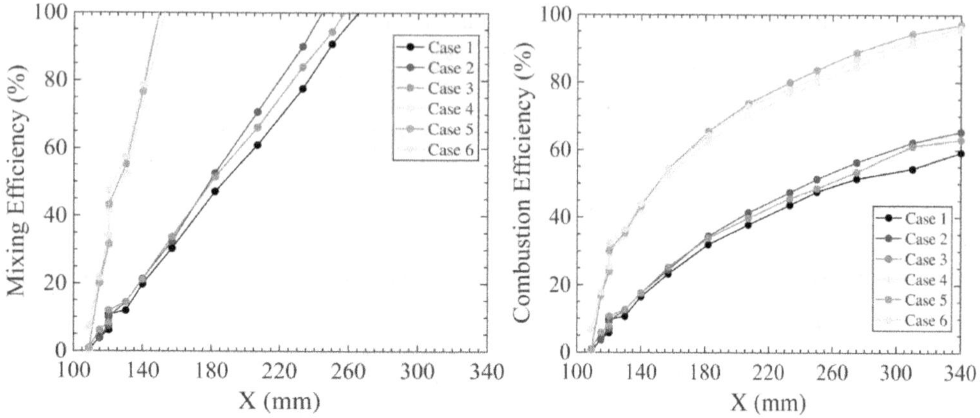

Figure 5. 13: Mixing Efficiency and Combustion Efficiency along the combustor length.

The present work simulates cavity and dual-strut configurations in a strut-based scramjet combustor by solving the 2D Reynolds-Averaged Navier-Stokes equations with combustion models using the finite rate/eddy dissipation chemistry model. The results yield some key findings:

1. The dual-strut configuration leads to more recirculation zones than the cavity design, thus greatly enhancing air-fuel mixing in the very limited time available for supersonic combustion.

2. Velocity profiles reveal that dual-strut cases are functioning at lower velocities, which makes possible more fuel-air interaction, as also confirmed by Mach contours that show relatively lower speeds in these configurations.

3. Examination of hydrogen mass fraction contours discloses that dual-strut cases have better mixing performance than the single one, since they have lower hydrogen mass fractions.

4. Comparison of H2O mass fraction and temperature profiles goes even further to show better combustion in dual-strut models, as being manifested by higher H2O levels and temperature values.

5. Efficiency calculations indicate that dual-strut configurations accomplish 100% mixing much earlier, with the length of the combustor being shortened by 40% and the combustion efficiency being increased by 33% compared to the standard DLR model.

The dual-strut arrangement is superior to the cavity one in that it achieves higher combustion efficiency and shorter combustor length, thus making it possible to generate more thrust. The cavity has a marginal impact, with only a 1% efficiency difference, and the cavity angle (45° vs. 90°) shows negligible effects on performance. While minor computational errors may exist, they are unlikely to affect the findings.

Future designs should prioritize the dual-strut configuration without a cavity to optimize performance and reduce complexity. Exploring configurations with a cavity on the upper wall or both walls could provide additional insights. Future studies can also evaluate the impact of cavity placement on flow characteristics and combustion performance in symmetrical or dual-cavity configurations. Future work that will incorporate three-dimensional, unsteady simulations to improve the accuracy and depth of analysis. Future work can also incorporate using multi-step reaction mechanisms to achieve higher fidelity results.

CHAPTER 6

INTEGRATED ANALYSIS OF SCRAMJET (INLET+COMBUSTOR)

6.1 INTRODUCTION

This chapter addresses the coupling of two very important parts of a scramjet engine: the combustor and the inlet. It is a continuation of the analyses that have been done in the preceding chapters, namely Chapter 4, which was on the scramjet inlet, and Chapter 5, which discusses about the scramjet combustor.

First, the chapter begins by setting the baseline models for the combustor and the inlet. The baseline models are used as a point of reference for subsequent modifications and optimizations. Subsequently, the research continues by combining these baseline models. This combination is crucial since it provides the platform for in-depth analysis on how the performance of the combustor and the inlet interact when combined. After the integration of the base models, the study continues to optimize every component separately. This includes implementing the best-case geometries realized for both the inlet and combustor into the integrated base case. Through this, we are trying to maximize the performance of every component independently.

Lastly, the chapter concludes with the integration of the optimized geometries of both the inlet and combustor into an integrated case. This integrated configuration is the culmination of the optimization process wherein the optimized designs of both components are combined. The performance of this fully integrated configuration is then completely evaluated afterwards. Evaluating the performance of the fully integrated system gives us the ability to determine how the optimized inlet and combustor designs complement each other and provide the overall efficiency and effectiveness of the scramjet engine.

Generally, this chapter is a well-structured methodology for optimization and integration of the major components in a scramjet engine for the purpose of maximizing its performance and capabilities

6.2 GEOMETRY:

Here, under the integrated base case geometry, attention is given to coordinating the dimensions of the inlet and combustor components for the attainment of smooth flow transition and efficient performance in the scramjet engine. For this, the combustor is scaled down proportionally so that its dimensions are equivalent to that of the inlet exit. For example, whereas the initial combustor inlet height was 50mm, it's now reduced to a much smaller 6.8mm to correspond to the inlet's exit height. This scaling guarantees that flow from the inlet runs smoothly into the combustor without going through severe changes in dimensions, thereby ensuring proper combustion processes and overall engine efficiency.

Options for Integrating the Two Geometries

Inlet exit height = 8.22 mm, Combustor inlet height = 50.00 mm.

Height ratio $\frac{\text{Hcombustor\_in}}{\text{Hinlet\_exit}} = \frac{50}{8.22}$ =6.08. So the combustor cross-section is 6.1× larger than the inlet exit.

Attaching them directly will cause severe mismatch and issues like choking, shocks, wrong residence time.

Option A: Match cross-sectional area at the interface

Case 1: Make inlet exit height= combustor entry height i.e 50.00 mm,

Linear scale for inlet: Inlet exit height =50.00/8.22=6.08.

Effect on inlet axial lengths (Inlet original axial segments 52, 30, 53 mm):

52 mm → 316.2 mm

30 mm → 182.4 mm

53 mm → 322.2 mm

New inlet total length= 820.8 mm

Drastically enlarges inlet: changes inlet aerodynamics (shock structure), wall-Reynolds numbers and computational cost.

Case 2: Make Combustor height to inlet exit height 8.22 mm then the combustor becomes very small which will affect the residence time, length of the combustor, mixing length, heat transfer, and combustion physics change drastically

Option B: Intermediate compromises scale

Scaling both parts to a target cross-section that is between the two extremes (reduces numerical stiffness). A good neutral choice is the geometric mean of the two heights.

Geometric mean height $H_t = \sqrt{Hcombustor_{inlet} * H\ inletexit} = \sqrt{50 * 8.22} = 20.25$ mm

(Combustor Inlet Height)

Inlet linear scale: Sin=$\frac{20.25}{8.22}$ = 2.46 (inlet enlarged moderately)

Combustor linear scale: Scomb = $\frac{20.25}{50}$ = 0.405 (combustor reduced moderately)

Numeric scaled dims:

Base Inlet (original lengths 52, 30, 53 mm, exit length=6.22):

52 → **154.85 mm**

30 → **89.34 mm**

53 → **157.83 mm**

new inlet total = **402**

inlet exit height → **20.25 mm**

Enhanced Inlet (original lengths 52, 30, 53 mm, exit length=8.22):

52 → **128.1 mm**

30 → **73.91 mm**

53 → **130.57 mm**

new inlet total = **332.57 mm**

inlet exit height → **20.25 mm**

Combustor (original length of combustor 340 mm; inlet height 50, exit height 62 mm):

340 → **137.7 mm**

combustor inlet height 50 → **20.25 mm**

combustor exit height 62 → **25.11 mm**

Total length of the Integrated Base case=540mm

Total length of the Integrated Enhanced case= 470 mm

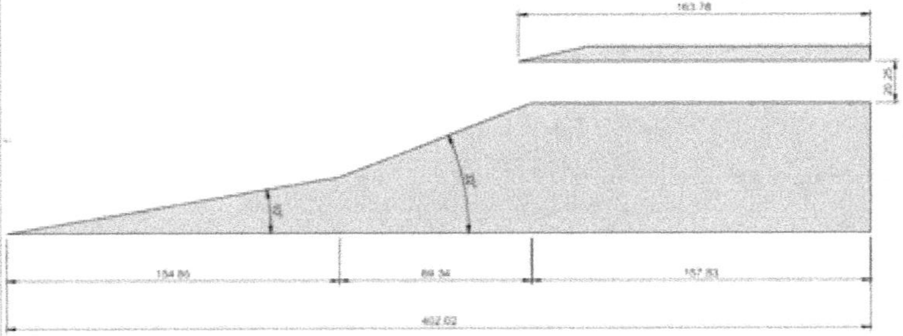

Fig. 6.1 a) Base Case of Inlet as per scaled dimensions

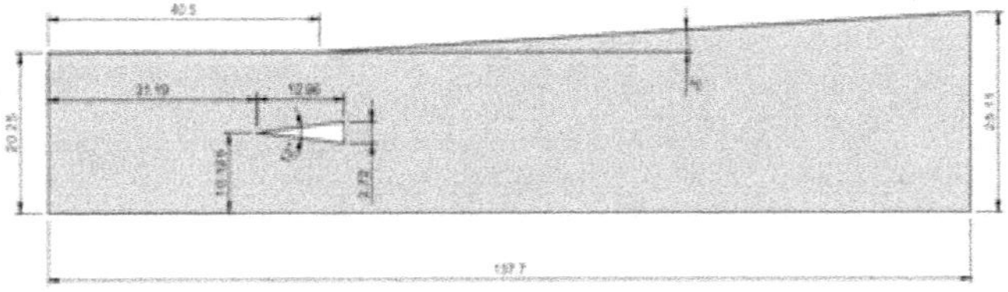

Fig. 6.1 b) Base Case of Combustor as per scaled dimensions

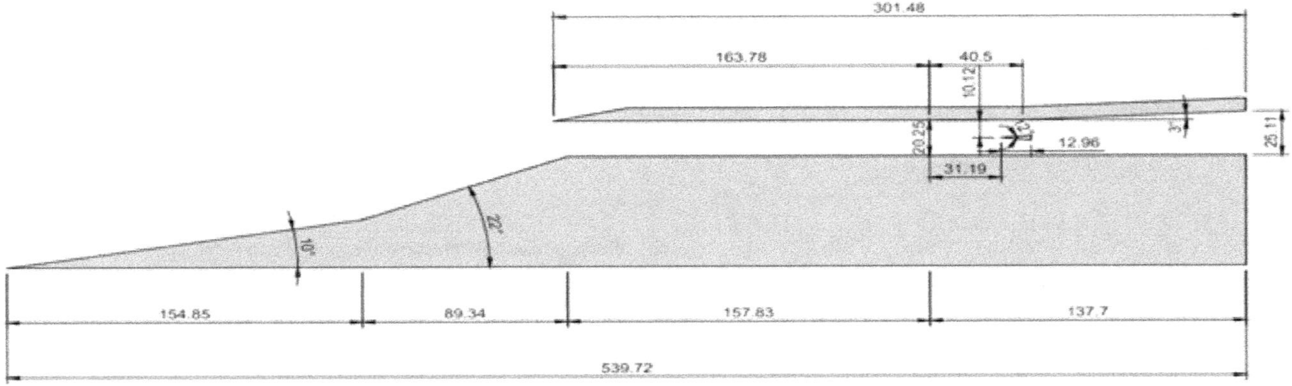

Fig. 6.1 c) Base Case of Integrated Inlet+Combustor (scaled dimensions)

Fig 6.1 Integrated base model of Inlet and Combustor

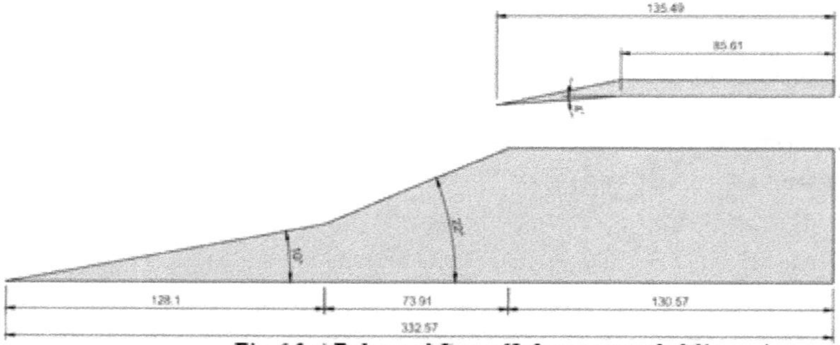

Fig. 6.2 a) Enhanced Case of Inlet as per scaled dimensions

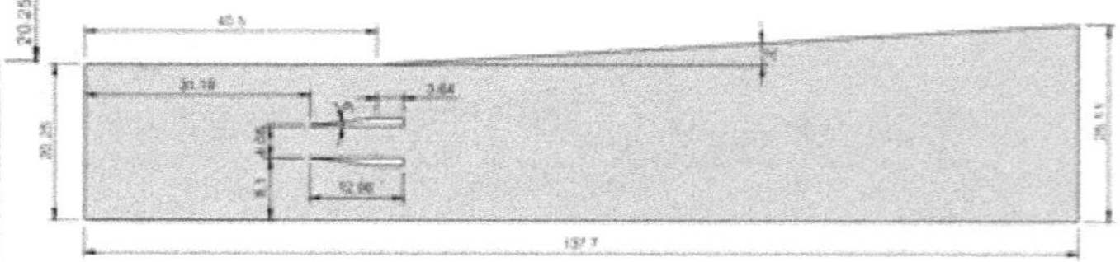

Fig. 6.2 b) Enhanced Case of Combustor as per scaled dimensions

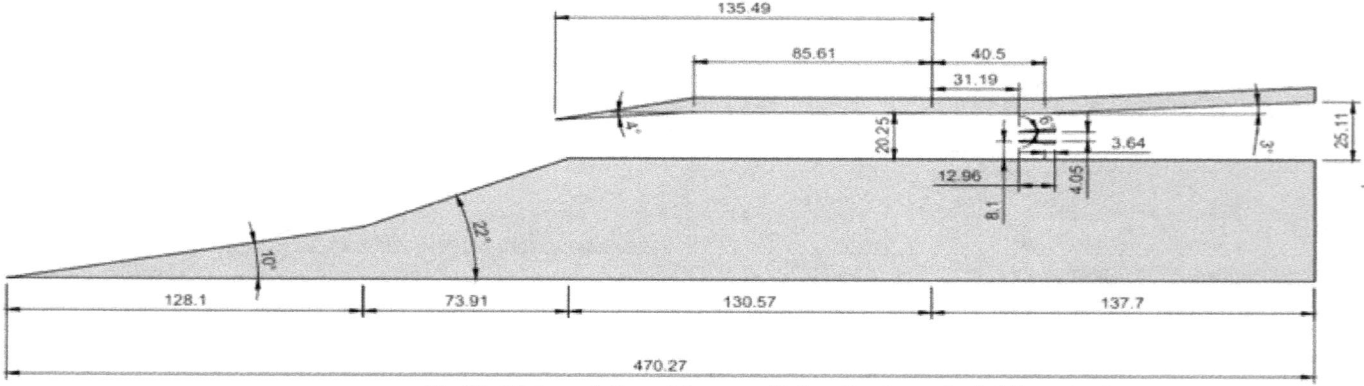

Fig. 6.2 c) Enhanced Case of Integrated Inlet+ Combustor (scaled dimensions)

Fig. 6.2 Enhanced Cases of Integrated Inlet and Combustor

6.3 DOMAIN AND BOUNDARY CONDITIONS:

The domain and boundary conditions adopted for further analysis is shown below in Fig 6.3 and Table 6.1. Meshing of coupled geometry base case is shown in Fig. 6.4

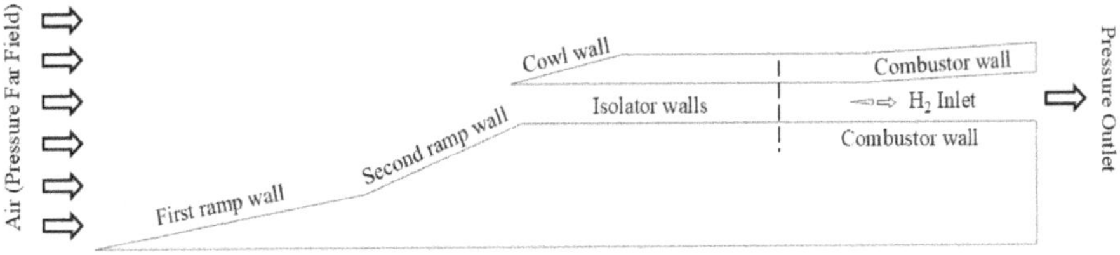

Fig 6.3 Geometry and domain of the integrated base case simulation

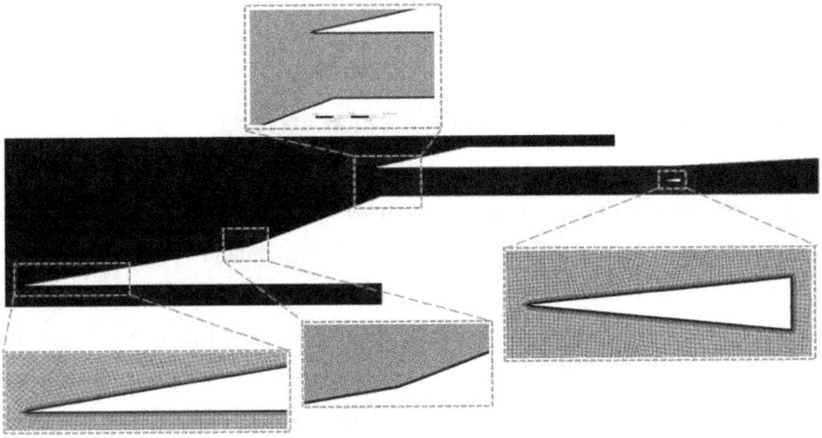

Fig 6.4: Meshing of coupled geometry base case.

Table 6.1 Boundary conditions for coupled geometry

Parameter	Air	H_2 (Fuel)
Boundary type	Pressure Far Field	Pressure Far Field
Mach Number	5	1
Velocity (m/s)	790.5793	1200
Static Temperature (K)	62.51	250
Density (Kg/m^3)	0.0707	0.097
Static Pressure (Pa)	1228.5	100,000
Y_{O2}	0.232	0
Y_{N2}	0.736	0
Y_{H2O}	0.032	0
Y_{H2}	0	1

6.4 PERFORMANCE OF INTEGRATED MODEL

6.4.1 Mach Contours

The Mach contours below provide a comprehensive qualitative investigation of the base and the enhanced integrated scramjet configurations. The contours with a base model inlet and combustor illustrate the fundamental flow features serving as a reference, thus, no enhancements are visible. Enhanced case commits to the integration of the optimized inlet and the combustor models, unveiling the synergistic effects and emphasizing the performance improvements due to the combined enhancements. These pictures open up a detailed evaluation of the aerodynamic performance, shock wave geometries, and flow interactions at different configurations, thus, helping to make informed design choices and promoting the advancement of scramjet technology.

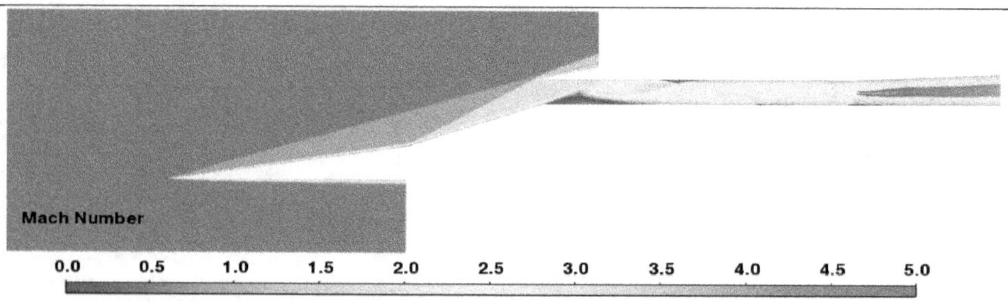

Fig 6.5 a) Mach Number counter of Base scramjet inlet and combustor

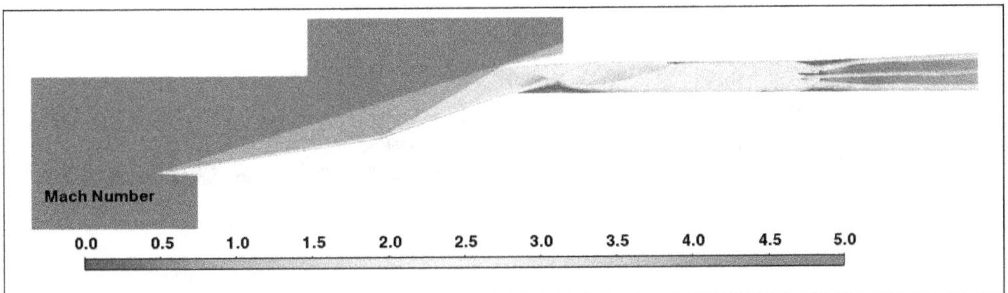

Fig 6.5 b) Mach Number counter of Enhanced scramjet inlet and combustor

6.4.2 Static Temperature Counters

The static temperature contours in Figure 6.6 (a-b) provide a visual representation of the Base and Enhanced scramjet inlet–combustor configurations, indicating the temperature distribution and combustion characteristics within the flow domain.

The Base scramjet has shown a lower temperature distribution throughout the combustor region with the static temperature values reaching up to around 2154 K. The regions of high temperature (shown as bright yellow–orange areas) are primarily confined to the flame region and remain relatively small in size, thus indicating limited heat release and combustion intensity. Also, the temperature gradient looks less uniform which signifies incomplete combustion and weaker mixing between fuel and air streams.

In contrast, the Enhanced scramjet reveals a substantially higher static temperature field with the temperature values going up to nearly 2890 K. The high-temperature region moves farther downstream, thus covering a larger part of the combustor and indicating stronger and more sustained combustion. The temperature distribution looks more uniform and intense, which reflects the improved mixing and flame stability that resulted from the applied enhancements (most probably geometric or flow-control modifications). In agreement with the first performance results, the higher exit temperature is indicative of more energy conversion taking place and therefore more thrust generation.

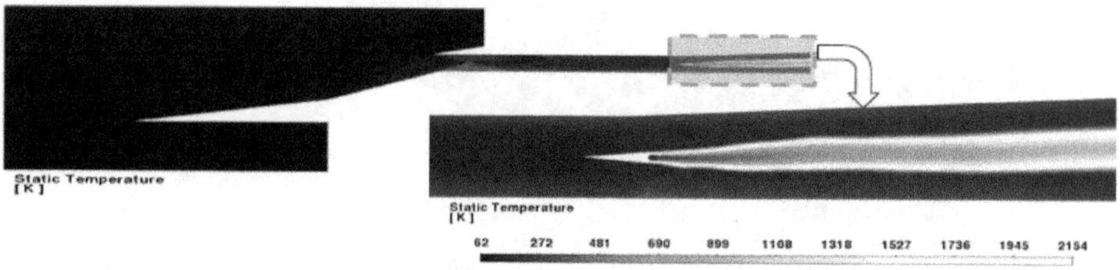

Fig 6.6 a) Static Temperature counter of Base scramjet inlet and combustor

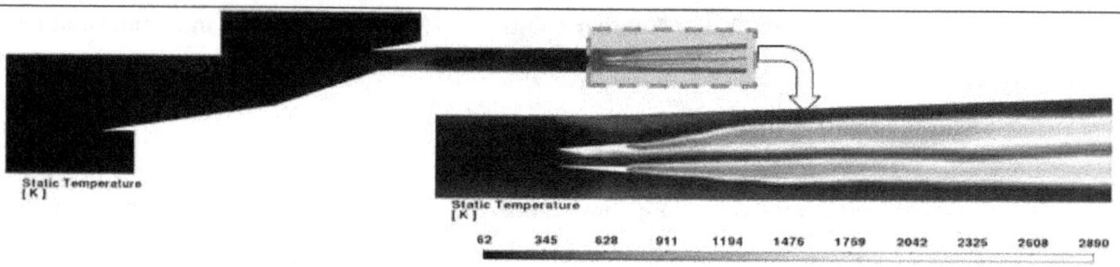

Fig 6.6 b) Static Temperature counter of Enhanced scramjet inlet and combustor

6.4.3 Mixing and Combustion efficiency

The efficiency with which the combustion process takes place in a Scramjet engine is a major factor determining the overall efficiency of the Scramjet engine. These numbers serve as a bridge for the improvements in performance resulting from the optimization of integrated scramjet configurations. It is apparent from the figure that, as both the inlet and combustor components are improved, the combustion efficiency shows a steady increase. There is a significant transition, moving from the base case to the enhanced case, which is a testament to the effectiveness of design refinements in promoting combustion processes within the scramjet engine. These efficiency figures are supported by the qualitative analysis that is given by Mach contours, thereby providing a complete insight into the performance gains accomplished by the different configurations.

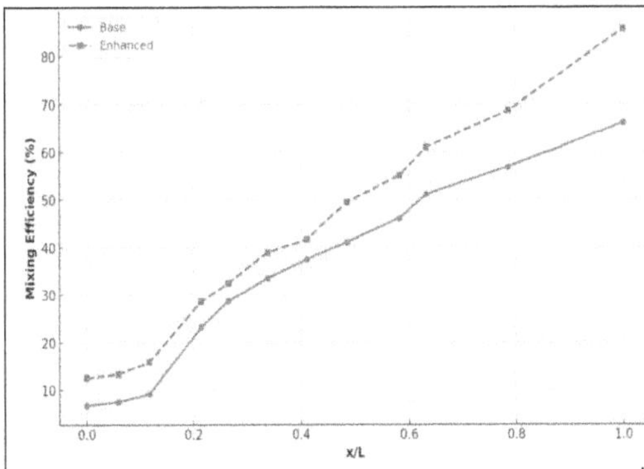

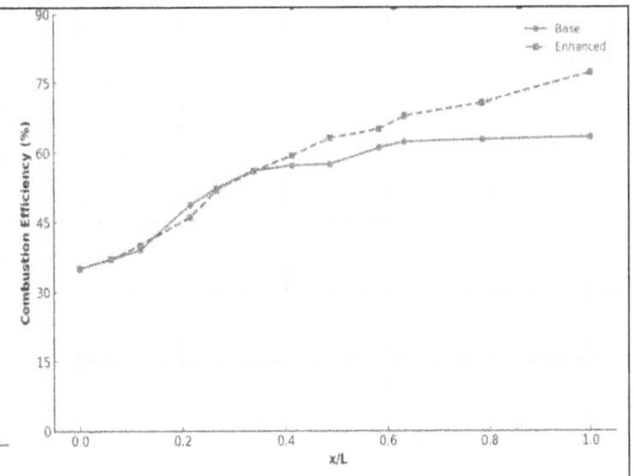

Fig 6.7 a) Mixing efficiency comparison for Base and Enhanced	**Fig 6.7 b) Combustion efficiency comparison for Base and Enhanced cases**

6.4.4 Different parameters comparison of integrated base and enhanced case

The Enhanced configuration is always better than the Base configuration in terms of performance metrics. The biggest improvement, by far, is Mixing Efficiency, while the other three parameters, namely Combustion Efficiency, Heat Release, and Flame Holding, have also progressed to a similar extent. This is evidence that the design changes in the Enhanced configuration lead to higher total combustor performance, which is the ultimate goal, namely, greater efficiency and stability. Figure 6.8 presents the performance of different parameters for the base and enhanced cases.

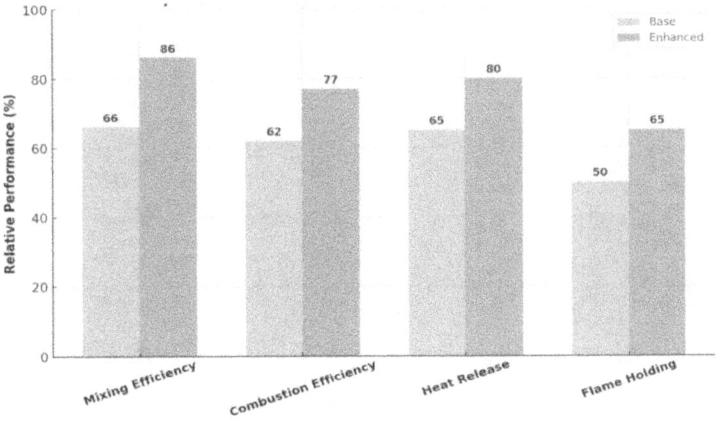

Fig 6.8: Comparison of combustor performance Base and Enhanced cases

6.4.5 Thrust and Specific Impulse for Base and Enhanced Cases

1. Thrust Comparison (Figure 6.9): The Base configuration produces a thrust of 24.42 kN, whereas the Enhanced configuration generates 26.83 kN. This means that the thrust of the enhanced case is about 9.9% higher than that of the base case. The increase in the net propulsive force due to the modifications in the enhanced configuration is the main reason for this improvement, which is probably a result of better combustion and heat release properties.

2. Specific Impulse Comparison (Figure 6.10): The Base configuration has a specific impulse of 1659.7 s, and the Enhanced configuration reaches 1823.6 s. The difference between these two values is approximately 9.9%, which indicates that the enhanced model yields higher propulsive efficiency. A higher specific impulse means that more thrust is generated per unit of fuel consumed, which is a clear indication of fuel utilization and overall engine efficiency improvement.

Both the thrust and the specific impulse have been significantly improved in the Enhanced configuration as compared to the Base model. It is evident from this that the scramjet system has achieved higher combustion efficiency, better energy conversion, and superior propulsive performance as a result of the design or flow-control modifications implemented.

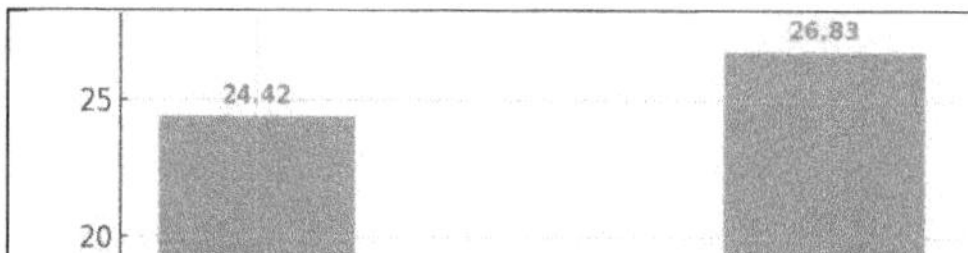

Fig 6.9 Thrust comparison for Base and Enhanced

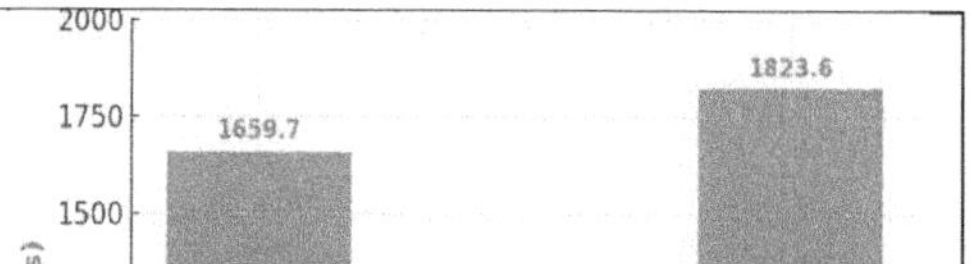

Fig 6.10 Specific Impulse comparison for Base and Enhanced cases

6.5 RESULTS AND DISCUSSION

The combined research on in-flight modification of the scramjet engine with the study of the rapid airflow intake system and the combustor unit is a main idea of this chapter, which deals with the method of enhancing performance and capabilities of scramjet engines. The very beginning of the journey was the grounding of the base model for both the intake and the combustor units, final stages – the evaluation of the completely integrated configurations. It all started with matching the sizes of the inlet and the combustor parts, and then changing the parts such as reducing the size of the combustor in order not only to ensure smooth flow transition but also to achieve the optimal engine performance. At the core of this painful operation is essentially a study set up for comparing different integrated model configurations, which is elaborated in the next sections.

The comparison between the base and integrated scenarios serves as a starting point for the identification of differences in the designs of the inlet and the combustor and the formulation of hypotheses about changes in overall engine performance as variables. The use of qualitative analysis with Mach contours aids the understanding of the flow phenomena, shock wave structures and aerodynamic interactions for different configurations in full scope. Moreover, the quantitative measurements of the combustion efficiency indicate that remarkable performance gains go hand in hand with the improvement of the inlet and the combustor design. The change from a combustion efficiency of 62% in the base case to 77% in the enhanced case is quite obvious, thus, the effectiveness of design refinements is demonstrated. This integrated strategy, which depends on both qualitative and quantitative analyses, not only extends our understanding of scramjet engine performance but also guides the further design and the development of future high-speed propulsion systems.

www.ingramcontent.com/pod-product-compliance
Lightning Source LLC
LaVergne TN
LVHW021736071126
851857LV00040B/1465